Muirhead Library of Philosophy

THE PHILOSOPHY OF WHITEHEAD

MUIRHEAD

Muirhead Library of Philosophy

METAPHYSICS
In 17 Volumes

I	Time and Free Will	*Bergson*
II	Reason and Analysis	*Blanshard*
III	Appearance and Reality	*Bradley*
IV	In Defence of Free Will	*Campbell*
V	Person and Object	*Chisholm*
VI	Non-Linguistic Philosophy	*Ewing*
VII	The Foundations of Metaphysics in Science	*Harris*
VIII	The Concept of Meaning	*Hill*
IX	Philosophy and Illusion	*Lazerowitz*
X	The Relevance of Whitehead	*Leclerc*
XI	Dialogues on Metaphysics	*Malebranche*
XII	The Philosophy of Whitehead	*Mays*
XIII	Studies in the Metaphysics of Bradley	*Saxena*
XIV	The Intelligible World	*Urban*
XV	Language and Reality	*Urban*
XVI	Valuation	*Urban*
XVII	Philosophy of Space and Time	*Whiteman*

THE PHILOSOPHY OF WHITEHEAD

W MAYS

Routledge
Taylor & Francis Group
LONDON AND NEW YORK

First published 1959 by George Allen & Unwin Ltd

Published 2013 by Routledge
2 Park Square, Milton Park, Abingdon, Oxfordshire OX14 4RN
711 Third Avenue, New York, NY 10017

First issued in paperback 2014

Routledge is an imprint of the Taylor and Francis Group, an informa business

The publishers have made every effort to contact authors/copyright holders of the works reprinted in the *Muirhead Library of Philosophy*. This has not been possible in every case, however, and we would welcome correspondence from those individuals/companies we have been unable to trace.

These reprints are taken from original copies of each book. In many cases the condition of these originals is not perfect. The publisher has gone to great lengths to ensure the quality of these reprints, but wishes to point out that certain characteristics of the original copies will, of necessity, be apparent in reprints thereof.

British Library Cataloguing in Publication Data
A CIP catalogue record for this book
is available from the British Library

The Philosophy of Whitehead
ISBN 978-0-415-29600-7 (hbk)
ISBN 978-1-138-87106-9 (pbk)
Metaphysics: 17 Volumes
ISBN 978-0-415-29532-1
Muirhead Library of Philosophy: 95 Volumes
ISBN 978-0-415-27897-3

MUIRHEAD LIBRARY OF PHILOSOPHY

AN admirable statement of the aims of the Library of Philosophy was provided by the first editor, the late Professor J. H. Muirhead, in his description of the original programme printed in Erdmann's *History of Philosophy* under the date 1890. This was slightly modified in subsequent volumes to take the form of the following statement:

'The Muirhead Library of Philosophy was designed as a contribution to the History of Modern Philosophy under the heads: first of different Schools of Thought—Sensationalist, Realist, Idealist, Intuitivist; secondly of different Subjects—Psychology, Ethics, Aesthetics, Political Philosophy, Theology. While much had been done in England in tracing the course of evolution in nature, history, economics, morals and religion, little had been done in tracing the development of thought on these subjects. Yet "the evolution of opinion is part of the whole evolution".

'By the co-operation of different writers in carrying out this plan it was hoped that a thoroughness and completeness of treatment, otherwise unattainable, might be secured. It was believed also that from writers mainly British and American fuller consideration of English Philosophy than it had hitherto received might be looked for. In the earlier of series books containing, among others, Bosanquet's *History of Aesthetic*, Pfleiderer's *Rational Theology since Kant*, Albee's *History of English Utilitarianism*, Bonar's *Philosophy and Political Economy*, Brett's *History of Psychology*, Ritchie's *Natural Rights*, these objects were to a large extent effected.

'In the meantime original work of a high order was being produced both in England and America by such writers as Bradley, Stout, Bertrand Russell, Baldwin, Urban, Montague, and others, and a new interest in foreign works, German, French, and Italian, which had either become classical or were attracting public attention, had developed. The scope of the Library thus became extended into something more international, and it is entering on the fifth decade of its existence in the hope that it may contribute to that mutual understanding between countries which is so pressing a need of the present time.'

The need which Professor Muirhead stressed is no less pressing today, and few will deny that philosophy has much to do with enabling us to meet it, although no one, least of all Muirhead himself, would regard that as the sole, or even the main, object of philosophy. As Professor Muirhead continues to lend the distinction of his name to the Library of Philosophy it seemed not inappropriate to allow him to recall us to these aims in his own words. The emphasis on the history of thought also seemed to me very timely; and the number of important works promised for the Library in the near future augur well for the continued fulfilment, in this and other ways, of the expectations of the original editor.

H. D. LEWIS

THE PHILOSOPHY OF WHITEHEAD

BY

W. MAYS

LONDON: GEORGE ALLEN & UNWIN LTD
NEW YORK: THE MACMILLAN COMPANY

FIRST PUBLISHED IN 1959

PRINTED IN GREAT BRITAIN
in 11 on 12-pt. Imprint type
BY C. TINLING AND CO., LTD.
LIVERPOOL, LONDON AND PRESCOT

The reformation of Logic has an essential reference to Metaphysics. For Logic prescribes the shapes of Metaphysical Thought.

WHITEHEAD: Forward to Quine's *System of Logistic*

PREFACE

IT is some years since I became interested in Whitehead. As I studied his writings, I became convinced that there was something wrong with the usual interpretations of them. As I read and re-read *Process and Reality*, it became increasingly clear to me that Whitehead was not quite the Platonist that he had been made out. This led me to a detailed examination of the book, relating it as far as possible to his earlier work. The late Professor Susan Stebbing had already emphasized in conversation with me the importance, for an understanding of Whitehead's philosophy, of his 1906 paper 'On Mathematical Concepts of the Material World'. This paper, in which Whitehead tries to apply mathematical logic to physical problems, strengthened my conviction that a new interpretation of Whitehead's philosophy was necessary.

Most of this work was done in the early years of the Second World War. After the war I put the manuscript on one side, having extensive teaching duties. My interests also moved in different directions. When I returned to the study of Whitehead a few years ago, I was able to see the work in a better perspective. I therefore decided to revise the manuscript and put it in book form. This entailed a certain amount of condensation and the deletion of some detailed textual commentary. I have also added some new material, for example, the chapter on 'The Method of Extensive Abstraction.'

The main purpose of this study is to illustrate the two principles upon which the Philosophy of Organism is based, (a) its logical or structural side, and (b) its physical-experiential side. Part I deals with (a) and Part II with (b). Each chapter in either part reflects some aspect or other of these two principles.

I have tried as far as possible to make a close study of Whitehead's text and not to indulge in generalizations which cannot be supported by the evidence. I have therefore at times had to give an almost word for word analysis of Whitehead's text. Further, it is not always possible to explain in ordinary everyday language Whitehead's philosophical concepts, since they often refer to technical mathematical and physical notions.

In brief, then, this book may be said to be a commentary on some of the more important aspects of *Process and Reality*. It should, however, be remembered that Whitehead's position is not always systematically developed in that work. I have therefore often had to examine in some detail his other writings, where his ideas are worked out at greater length and in a more cogent fashion.

I owe thanks to Professor D. M. Emmet, Dr. A. C. Ewing and Professor H. D. Lewis, all of whom made valuable suggestions for the improvement of the book. I am greatly indebted to Dr. David Rees and Miss Clare Oxburgh for the great care and patience with which they have read the proofs, and also to Miss P. Horne for typing the manuscript.

Chapters I and IV were originally printed in the *Proceedings of the Aristotelian Society* of 1945 and 1952, respectively, and Chapter XVII in *Philosophy and Phenomenological Research*, Vol. XV, 1955.

December 1st, 1958. W. MAYS

CONTENTS

INTRODUCTION

DURING the last decade interest in Whitehead's work has been slight. This is partly due to the anti-metaphysical trend of modern philosophy and partly to the extreme difficulty of understanding his later work. It is sometimes overlooked, because of this, that he was the co-author of *Principia Mathematica*, and that the central doctrines of his later metaphysical writings have their origin in his preoccupation with mathematical logic and its applications. In recent years, especially in the United States, there has been some resurgence of interest in his writings, and even positivist reviewers have noted their modern ring. The reason is not far to seek. At a number of points Whitehead's philosophy touches upon problems which are being discussed nowadays, if not by philosophers, at least by cyberneticians, psychologists and philosophers of science. He was indeed one of the first to apply symbolic logic to concrete problems.

What has been more and more demonstrated is that there is greater uniformity in his thought than has usually been assumed. In view of this it may be worth while giving some outline of Whitehead's philosophical doctrines, bringing out the relationship between his earlier more scientific philosophy and his later metaphysics. Our account may seem very condensed, but the purpose of this introduction is to indicate certain points which are of interest, to be treated at greater length later. Three periods may be discerned in his work, (i) that in which he busied himself with mathematics and mathematical logic, (ii) the period of his nature philosophy, in which he seemed to be putting forward a *Logische Aufbau der Welt*, showing how mathematical and physical concepts could be derived from experienced events, and (iii) his metaphysical period which largely begins with his translation to Harvard. It is the last which has aroused the most criticism and led him to be neglected by the more tough-minded philosophers.

A cursory examination of *Process and Reality* is sufficient to show that it is one of the most difficult philosophical works written in the English language. The reader is faced with a formidable catalogue of such unfamiliar expressions as the Primordial Nature of God, the realm of eternal objects, conceptual and physical prehensions, subjective aim and concrescence—a vocabulary which seems shot through with subjectivism and is calculated to send a shudder down the spine of the semantically sophisticated philosopher. The whole picture seems very alien to his earlier writings, yet a closer examination will show that, despite the formidable terminology, the ideas contained in his later work are much simpler than is usually assumed, since he is working out some of his earlier ideas on a larger philosophical canvas.

In his earlier philosophy, developed in *The Concept of Nature* and *The Principles of Natural Knowledge*, Whitehead was mainly concerned with accepting the results of modern physics: physical nature as distinct from the immediately perceived events was taken as a four-dimensional structure of events pervaded by electromagnetic characteristics. Whitehead was not therefore simply putting forward a phenomenalism; he believed that the world described by the physicist really did exist, though perhaps not quite in the neat form in which it appeared in the physicist's equations. It is here that the Method of Extensive Abstraction proved its value, since as a logical instrument it enabled him to relate such precise geometrical concepts as volumes, straight lines and points, conceived in the form of ideal convergent series, to sensory events. His early philosophy of nature therefore contains a *Logische Aufbau*, but one giving an independent status to the world of physics. A feature of this work as seen, for example, in *The Principle of Relativity*, was that he put forward his own brand of relativity theory, in which, unlike Einstein, he did not take physical space-time as heteroloidal, since he regarded it as having a uniform structure similar to perceptual space-time. Whitehead argued that perceptual space-time had to be uniform to allow for judgments of congruence upon which measurement was based, and that in order that induction should be possible the relations within our perceptual field had to be identical with those outside it.

At a first glance the philosophical system elaborated in *Process and Reality* seems a radical departure from his early writings. But when we examine what Whitehead means by speculative philosophy and his account of philosophical method, we find that it bears a close resemblance to the axiomatic method used by modern logic. However, Whitehead makes little attempt to formalize his account, except for one section in which he discusses the relationship of extensive connection (defined in terms of overlapping, etc.) holding between abstract topological regions. He further conceives these general topological relationships as forming an abstract system of order underlying the universe, and defining an infinite range of possible space-time orders, the present universe being regarded as one such.

Against this background Whitehead works out his philosophy. Two kinds of entities are postulated, those making up the general scheme of order (or God) underlying the universe, and which guarantee induction, and the physical events (or the World) related within this structure. Further, he contends that the world of physics, described in terms of the transference of energy in the electromagnetic field, has analogous properties to the sensory and emotional elements immediately experienced by us, as he describes the physical quanta of energy as primitive throbs of emotional intensity. For Whitehead the world is made up of co-ordinated systems of events or societies, some of which give rise to complex wholes manifesting new properties—life and mind, for example, are considered to be Gestalt properties of such systems of events. And since in his view psychological systems have in some respects similar properties to physical ones, there can, he assumes, be a causal interaction between them.

In his early writings Whitehead tended to gloss over the precise nature of the mechanisms involved in our perception of the external world, which at first sight made it appear that he was putting forward a variety of phenomenalism. In his later work he is considerably more explicit, since he asserts that the experienced sensory qualities are obtained by a process of statistical averaging from the physiological and physical activities involved in perception—a process which seems to resemble that postulated by some cyberneticians in their attempt to explain pattern recognition. Whitehead further believes that in our ordinary everyday life we use our sensory data as symbols for the actual events in physical nature. We are enabled to do this, he argues, since (1) they both partake in the same general scheme of order, and (2) our sensory data are causally related to the physical events in the immediate past which have given rise to them. Whitehead also attempts to overcome Hume's objection that causality is not a perceptual quality by postulating a special kind of causal perception. This, he asserts, gives us direct awareness of the functionings of our sense-organs and of the surrounding efficacious physical world; an awareness which he claims is more fundamental than that given by the specific colours, sounds, etc., in our perceptual field, which only have an indirect reference to the physical events.

Though it is obvious that logic and mathematics play an

important part in Whitehead's system, the part played by physical concepts has usually been overlooked. Despite Whitehead's claim that his account of perception is closer to the perceptual facts than that of other philosophers, he seems nevertheless to use physical field theory, as for that matter do the Gestalt psychologists, as a model for his descriptions of experience. When Whitehead therefore asserts that he is putting forward a philosophy of organism, this should not be taken to mean that he bases physics on biology, but merely that he believes they both deal with systems having a historicity about them. He is pointing out that in the physical world as well as in human experience the way things develop is determined not only by their present situation but by their whole past history. It is this which gives his philosophy a Hegelian tinge. And, as we shall see, the two key notions of Whitehead's later philosophy are the postulational method of modern logic with its emphasis on complex relational systems, and the field theory of modern physics with its emphasis on the historicity of physical systems.

Looked at in this way, Whitehead's account does not seem to be as outrageous nor as metaphysical as some philosophers have made it out to be, since what he seems to be doing is a sort of applied logic. Even if one refuses to regard this kind of activity as philosophy, nevertheless there are today similar lines of enquiry, for example, the more speculative sides of cybernetics, though admittedly cyberneticians express their speculations in mathematical formulae rather than philosophical language.

Of considerable importance for an understanding of Whitehead's work is his Royal Society paper 'On Mathematical Concepts of the Material World'. Whitehead here attempts to apply the axiomatic method to physical science. It is interesting to note that his account of a many-termed relation and its field, the entities related by it—in this case physical lines of force—bears a strong similarity to his account in *Process and Reality* of a general system of order permeated by electro-magnetic events. The change in Whitehead's view would not, then, seem to be as radical as has usually been supposed, since in some ways he might be said to be returning to his 1906 position. The Royal Society paper probably gives a much clearer insight into Whitehead's later philosophy than the many analogies that have been drawn between his work and such philosophers as Plato.

If I have any criticism to make of Whitehead's later position it is that I am sceptical of any attempt, philosophical or otherwise, to regard the universe as a logical system. I am also somewhat unhappy about Whitehead's identification of physical field theory with the experienced passage of nature; the former seems to be more of an abstract mathematical construction than an experienced fact. But nevertheless, it needs to be pointed out that quite a number of physicists would disagree with the phenomenalist approach and accept the reality of physical entities, though not perhaps Whitehead's account in terms of emotional energy.

ABBREVIATIONS

AE	*The Aims of Education and other Essays*	1929
AI	*Adventures of Ideas*	1933
CN	*The Concept of Nature*	1920
MC	'On Mathematical Concepts of the Material World', *Phil. Trans. Roy. Soc. A.* Reprinted in *Alfred North Whitehead: An Anthology*, 1953.	1906
MT	*Modes of Thought*	1938
NL	*Nature and Life*	1934
PNK	*An Enquiry Concerning the Principles of Natural Knowledge*	1919
PR	*Process and Reality*	1929
P of R	*The Principle of Relativity with applications to Physical Science*	1922
RM	*Religion in the Making*	1926
SMW	*Science and the Modern World*	1926
UA	*A Treatise on Universal Algebra*	1898

All the above are published by the Cambridge University Press, except the *Phil. Trans. Roy. Soc.*, and the *Aims of Education*, which is by Williams and Norgate. In the page-references to the text the following re-issue has been used, *Science and the Modern World* 1929. Where both English and American editions exist, the reference is always to the English edition.

PART ONE

THE STRUCTURE OF WHITEHEAD'S PHILOSOPHY

ABSTRACT

IN this part an examination is made of the formal side of Whitehead's philosophy. He analyses the universe of events into (i) the logical framework of order, and (ii) the temporal process which transforms it into the concrete spatio-temporal structure of events. There would seem to be a similarity between Whitehead's philosophical method and the hypothetico-deductive method of science. Whitehead contends that a philosophic scheme is to be verified in the same way as any other hypothesis, namely by its general success in practice. He also argues that there is no precise correspondence between ordinary language and experience. An account is given of Whitehead's conception of Deity as the ordering entity in nature, as well as of his philosophy of religion. According to Whitehead, the particular form taken by a religion is dependent upon our psychological and social reaction to these elements of order in nature. Further, he believes that owing to the complete generality of symbolic logic, it can be applied to fields which so far have not been found amenable to quantitative methods. With this in mind, he suggests that symbolic logic may yet become the foundation of aesthetics, ethics and theology. By the realm of eternal objects, Whitehead really refers to the logical form or abstract framework of the spatio-temporal structure of events, which in itself defines an infinite range of possibilities. Whitehead seems to be giving here a mathematical analysis of experience in place of the Aristotelian classificatory analysis. Finally the Method of Extensive Abstraction is discussed. It is suggested that the 'method' is not really an attempt to derive geometry from sensory data, but is to be considered after the fashion of an abstract mathematical model.

CHAPTER I

SPECULATIVE PHILOSOPHY

I

WHITEHEAD stands out among modern philosophers by his wide range of interests and breadth of vision. His ideas have influenced thinkers in such diverse fields as literature, biology and mathematics. Yet in philosophy his reputation has somewhat suffered, partly as a result of his obscurity of expression, which has made a judicious appraisal of his work very difficult. Despite these stylistic imperfections, even philosophers who are suspicious of any form of metaphysics would admit that there is something of importance in his philosophy, but what it is they are not very certain. It is therefore worth while giving some account of the basic conceptions of his philosophy. This we think may best be done by examining fairly closely what Whitehead conceives to be the true function of philosophical thinking.

Whitehead puts forward what he considers to be the true method of philosophic construction in his preface to P R. We are told that it consists in framing the best scheme of ideas that we can, and unflinchingly exploring the interpretation of experience in terms of it. There is a familiar ring about this and we do not have to go very far to find the reason. The method described above resembles that used by science, where from the particular observed data one frames the best possible theory and proceeds to see how it works out in practice by applying it to other facts. In effect, he would seem to use the hypothetico-deductive method to help in the elucidation of philosophical problems.

Whitehead makes it clear that the constructive procedure whereby we attempt to formulate a general scheme of which the particular facts are instances, is a method owned in common by both metaphysics and science.[1] He thereby emphasizes that his approach to philosophy resembles that of the scientist in his own special field of study. For example, note how in physics we endeavour to

[1] P R, cf. p. 57.

obtain more and more general theories and laws; from the falling bodies of Galileo and Kepler's laws of motion to Newton's more general scheme which places them under one heading as examples of the general theory of gravitation. So in the same way Whitehead is trying to find a scheme of the highest order of generality made up of more general notions than those found in any of the sciences—notions which are applicable to every kind of experience.

Let us now look at the way Whitehead defines 'speculative philosophy' in PR. 'Speculative Philosophy', he tells us, 'is the endeavour to frame a coherent, logical, necessary system of general ideas in terms of which every element of our experience can be interpreted.' That is to say, 'everything of which we are conscious, as enjoyed, perceived, willed, or thought, shall have the character of a particular instance of the general scheme'.[1]

At the outset it is well to state a difficulty. It might seem that the coherent, logical, necessary system Whitehead is urging us to put forward, and in terms of which all experience is to be interpreted, is the system of categories elaborated in PR, pp. 24-41. There would be some reason for supposing this if Whitehead simply meant by the categories the logical conditions to which all possible experience must conform. But it appears that by the 'Category of the Ultimate', i.e. Creativity (which expresses the general principle presupposed in the more special categories),[2] he is referring to something having definite empirical characteristics and resembling his earlier view of temporal process.

Now if this notion of creativity is to be taken as a metaphysical first principle, how can it be spoken of as coherent, logical and necessary, since in its essence it seems to have a temporal character? Such an interpretation would also run counter to what Whitehead means by a metaphysical principle in other parts of his work, where this is connected with mathematical pattern, e.g. his view that 'Poetry allies itself to metre, philosophy to mathematic pattern'.[3] How, then, are we to understand this? Do metaphysical first principles deal with empirical activities or do they deal with logico-mathematical principles? And the question is extremely difficult to decide since the evidence points either way.

We are not the first to have been puzzled by this dilemma. Dewey, after quoting the above definition of speculative philosophy, points out that it is suggestive of the kind of structure exhibited

[1] PR, p. 3. [2] PR, loc. cit., cf. p. 27. [3] MT, p. 238.

in pure mathematics. 'It seems to go much further than the mere statement—to which no exception can be taken—that the different portions of any philosophic scheme must hang together. For it makes, if I understand it aright, an assertion about what the constituents of nature itself *must* be in and of themselves.' Dewey goes on to state that this approach would seem to accord with the assimilation of the proper subject matter of philosophy (the constitution of nature) to that of mathematical theory. He concludes: 'Deficiency of my own intellectual grasp may be the cause of my belief that this entire strain of thought substitutes abstract logical connectedness for the concrete existential *temporal* connectedness upon which I have based my interpretation of Whitehead's system'.[1]

If we merely glance for a moment at S M W it will be seen that there is some ground for Dewey's remarks. We find Whitehead asserting that the harmony of logic lies upon the universe as an iron necessity.[2] When speaking, for example, of the abstract patterns of relationships investigated by mathematics, he even goes so far as to say they are also imposed on external reality; that 'This is nothing else than the necessity of abstract logic which is the presupposition involved in the very fact of interrelated existence'. And he considers this related togetherness to be 'the primary article of metaphysical doctrine'.

Now what seems to emerge is that Whitehead definitely thought of nature after the manner of a coherent system. He believed that the community of occasions is bound together by a logical framework which approximates to the sort of structure one investigates in mathematics. This at least appears to follow from his assumption in S M W that the necessities of abstract logic are involved in the very fact of interrelated existence. Dewey's criticism would then seem to be to the point. 'In this case, it is not simply the philosopher who must proceed logically. The scheme of nature and immediate experience is itself a logical system—when we have the wit to make it out in its own terms.'[3]

However, nature is not merely such an abstract system. Whitehead emphasizes that in our actual world the scheme is given a content by the creative advance of nature. In other words, we deal with concrete events illustrated by specific characteristics and

[1] *Philosophy of A. N. Whitehead*, ed. P. A. Schilpp, pp. 657-8.
[2] S M W, cf. p. 24.
[3] *Philosophical Review* 1937, p. 174.

related together in one spatio-temporal structure. It would therefore seem that the metaphysical situation (the actual world) to which our speculative philosophy must apply has two sides to it —(a) the logical framework of order, and (b) the temporal process which actualizes it (or gives it a factual content), transforming the abstract scheme into the spatio-temporal structure of events. A comparison might perhaps be drawn with his earlier view. 'There are two sides to nature, as it were, antagonistic the one to the other, and yet each essential. The one side is development in creative advance, the essential becomingness of nature. The other side is the permanence of things.'[1]

The difficulty Dewey gets into arises then from the fact that he considers one aspect of Whitehead's system at a time. However, not only have we the historic process of the world; it is also permeated by a framework of order which binds the events of nature into a coherent system. Whitehead's answer to Dewey's question, then, is that both principles—temporal process and logical order—are essential for our understanding of nature. If we consider them in abstraction from each other we shall get a one-sided view of things.

Nevertheless, we do agree with Dewey that when Whitehead in his definition of speculative philosophy speaks of it as an attempt to frame a coherent, logical and necessary system of general ideas, he would seem to be taking as a model the purely abstract systems one comes across in mathematics and formal logic. He speaks of a logical framework exhibiting a definite form of order.

As far as the word 'interpretation' is concerned, this term has a number of distinct senses. It might perhaps be assumed that Whitehead means by it what some philosophers have meant in the past—i.e. to start from a number of simple elements, say space-time, as, for example, Samuel Alexander does, and to attempt to show how the complexity of the universe can be built up out of them. Whitehead, however, would seem to be using the notion of interpretation in a rather different manner, in the way it is used in symbolic logic—where an abstract system is spoken of as being given an interpretation by definite values. In this case the values may be identified with our individual experiences, which are the particular instances ordered within our concrete spatio-temporal scheme, and which may be conceived as a specific interpretation of this abstract logical framework of order.

[1] P N K, p. 98.

It is clear that by 'interpretation' Whitehead meant at least the correlation of particular experiences in terms of such a scheme, for he tells us: 'there are no brute, self-contained matters of fact, capable of being understood apart from interpretation as an element in a system'.[1]

But what also emerges is that to interpret an instance of experience does not mean explaining it in terms of certain basic principles, but rather consists in relating it as an element in the spatio-temporal system of things. This is further borne out by his statement: 'Thus the understanding of the immediate brute fact requires its metaphysical interpretation as an item in a world with some systematic relation to it'.[2] But, it might be said, if the above is really the case how can you justify your own assertion that Whitehead merely meant by this notion the giving of a factual content to an abstract system? We need, however, to point out that when we deal with the extensive scheme (i.e. the logical framework) in relation to particular experiences it becomes the spatio-temporal structure of events. These experiences have therefore to be expressed in terms of this more concrete structure.

According to Whitehead, if we wish to understand any individual matter of fact we require to know that it is systematically related to elements other than itself in the past, present and future. Hence he believes that, unless there was such a framework of order pervading nature, systematic knowledge would be impossible, since we would not be able to compare our experiences or set up any kind of systematic relation between them. As he puts it, 'It is by reason of this disclosure of ultimate system that an intellectual comprehension of the physical universe is possible. There is a systematic framework permeating all relevant fact. By reference to this framework the variant, various, vagrant, evanescent details of the abundant world can have their mutual relations exhibited by their correlation to the common terms of a universal system.'[3] It would, then, seem that the correlation of these details in terms of a universal system comes to much the same thing as their interpretation as elements in such a scheme. We are thus given some ground for believing that this systematic framework is the coherent, logical and necessary system Whitehead is talking about in his definition of speculative philosophy.

When Whitehead, therefore, asserts: 'Here "interpretation"

[1] P R, p. 19. [2] P R, p. 19. [3] P R, p. 464.

means that each element shall have the character of a particular instance of the general scheme',[1] he is really saying that every element of our experience shall be related within such a systematic framework. This is very like his statement: 'to be an actual occasion in the physical world means that the entity in question is a relatum in this scheme of extensive connection'.[2] Whitehead believes that extensive connection is the primary relation holding between events. In other words, the events in this scheme have certain common relationships of whole and part, overlapping, etc. Further, he seems to think that some of the general characteristics of extension have probably an ultimate metaphysical character, persistent in every cosmic epoch,[3] i.e. they determine conditions to which all events have to conform. Evidently Whitehead considers these extensive relations to be more fundamental than the empirical relata, for he seems to hold the view that physical space and time and the general notion of creative advance presuppose the more general relationship of extension.

It should be noted that when Whitehead speaks of metaphysics in its proper general sense[4] (or of metaphysical necessity), he has in mind this abstract scheme of extensive relations which, owing to its generality, applies to every cosmic epoch. On the other hand, when he considers cosmology he deals with empirical elements, with our particular cosmic epoch characterized by electro-magnetic occasions, dimensions, shapes and measurability, which are not, however, metaphysically necessary. Other epochs might well have different types of order and characteristics. Hence, though the cosmological scheme refers to our actual world, metaphysics as thus defined concerns itself only with the most general conditions underlying the universe.

In the light of the above discussion we might perhaps say that the categories listed in PR, Part I, Chapter II, are empirical principles applicable to our particular cosmic epoch and in terms of which the 'Theory of Prehensions', or Whitehead's cosmological scheme, is to be elaborated.

We have, however, also to take into account the 'Theory of Extension', in which he discusses the characteristics of this logical scheme—the generic form pervading all experience. Indeed, he conceives the extensive continuum as an abstract system, and puts forward a set of postulates in terms of which its basic notions may

[1] AI, p. 285. [2] PR, p. 409. [3] PR, cf. p. 408.
[4] PR, cf. p. 126.

be defined.[1] And insofar as the extensive scheme is based on a set of postulates it is in itself a hypothetico-deductive system and hence embodies the method of the working hypothesis.

II

We need now to enquire what Whitehead means when he tells us that the philosophic scheme has to be coherent. Apparently by 'coherent' he means 'that the fundamental ideas, in terms of which the scheme is developed, presuppose each other so that in isolation they are meaningless'. Now he is not merely saying that the propositions in terms of which the philosophic scheme is expressed must hang together, but that the entities in the universe (or cosmological scheme) should be systematically related to one another. For he goes on, 'it is presupposed that no entity can be conceived in complete abstraction from the system of the universe, and that it is the business of speculative philosophy to exhibit this truth. This character is its coherence.'[2] By the coherence of the philosophic scheme Whitehead then means that no entity can be abstracted from its relationship to other entities within the coherent system which is the universe.

But the notion of coherence as used in the context of PR is by no means new; speaking of events and the space-time continuum, Whitehead pointed out that we must get rid of the false idea that nature is a mere aggregate of independent entities each capable of isolation.[3] And this seems very much like the truth which he considers it is the business of speculative philosophy to exhibit. There is evidently some connection with the doctrine of the uniform significance of events elaborated in his earlier works: i.e. that each event presupposes the rest of the uniform spatio-temporal structure in which it is related as an element. To give an illustration, we are told 'There is no such thing as crimson lone and by itself apart from nature. . . . You may put it this way, nature as a

[1] PR, cf. Part IV, Chapter II, Section II. Though the set of postulates for extensive connection has the character of a deductive system, the four types of categories postulated by him in PR, the categories of the ultimate, of existence, of explanation and the categorial obligations do not possess this character. They are rather classificatory principles descriptive of empirical processes. Unlike the principles of extensive connection they do not hold necessarily for every cosmic epoch.

[1] PR, p. 3. [2] CN, cf. p. 141.

system is presupposed in the crimsonness of the cloud'.[1] That is to say, whenever we perceive a particular colour or sound, etc., it is always related to a systematic background of fact.

The notion of coherence as defined by Whitehead would seem to find its most general application in the case of an abstract system where every proposition stands in relation to every other.[2] But he takes a further step and goes on to assume that the universe is also modelled on these lines; that each event presupposes every other. Whitehead states, for example, that 'the philosophical scheme should be coherent, logical, and, in respect to its interpretation, applicable and adequate.'[3] By this he would appear to mean that its formal side should be (i) coherent—every entity should be connected within the system—and (ii) logically consistent with other entities in the same scheme. As regards its empirical interpretation, the scheme should be (iii) applicable at least to some items of experience—and (iv) adequate. There should be no items incapable of being related as elements in the spatio-temporal system.

Whitehead further amplifies the above by saying, 'this ideal of speculative philosophy has its rational side and its empirical side'.[4] That is to say, the philosophic (or cosmological) scheme has its rational or formal side—the logical systematic framework of extensive relations—whilst the empirical side deals with the way this scheme applies to observed experience in the shape of the spatio-temporal system of related experiences. But, he goes on, the empirical elements and the relational framework are bound together, for 'The adequacy of the scheme . . . means that the texture of observed experience, as illustrating the philosophic scheme, is such that all related experience must exhibit the same texture'.[5] In other words, as a result of the scheme comprehending the whole variety of experience, every item has to appear as related within the limits of its range. Further, it means that the particular spatio-temporal relationships in observed experience are, to use the language of C N, 'uniformly significant'. They are the same beyond our immediate field as they are within it.

[1] *Proceedings of the Aristotelian Society*, 1921-2, p. 219.

[2] Stebbing tells us (*A Modern Introduction to Logic*, 5th ed., p. 198), 'A system is said to be *coherent* if *every* fact in the system is related to every other fact in the system by relations that are not merely conjunctive. A deductive system affords a good example of a coherent system.'

[3] P R, p. 3. [4] P R, p. 4. [5] P R, p. 4.

III

Philosophers, Whitehead tells us, can never hope finally to formulate the metaphysical first principles.[1] The reason for this lies partly in the weakness of our imaginative insight and partly in the deficiency of our everyday language, which is not of sufficient generality to express these ideas. With regard to the latter difficulty it might, however, be asked whether, if our language is not general enough, we could find a way out by the use of a suitable logical language. This, it may be thought, should enable us to deal with most types of generality, especially the sort of structure (patterns of connection[2]) investigated by mathematical logic, with which he thinks it should be the business of metaphysics to deal.

Whitehead probably had such an approach in mind, for he is of the opinion that the method of algebra (or symbolic logic) 'embodies the greatest discovery for the partial remedy of defective language',[3] since algebra endeavours to exemplify in its written structures the conceptual pattern which it wishes to convey. Hence, unlike our everyday language, which has been designed to handle the ordinary things of life and therefore lacks generality, the algebraic language enables us to represent the most general conditions metaphysics seeks in experience.

Further, owing to the weakness of our imaginative penetration, we have difficulty in comprehending these first principles apart from the particular experiences in which they are exemplified. That is to say, we are unable to obtain a direct insight into the logical framework of order pervading the universe, since it is, as it were, shrouded from our observation by the particular matters of fact. Whitehead elaborates on this by saying that, putting aside the difficulties due to the limitation of language, 'deficiency in imaginative penetration forbids progress in any form other than that of an asymptotic approach to a scheme of principles, only definable in terms of the ideal which they should satisfy'.[4] By which he would seem to mean that we can arrive at this scheme of principles only as a result of a process of approximation analogous to that employed, for example, in the Method of Extensive Abstraction, where we substitute routes of approximation to an ideal limit for the crude data observed in direct experience.

[1] PR, p. 4. [2] Cf. AI, p. 197.
[3] Cf. *Philosophical Review*, 1937, p. 183. [4] PR, p. 5.

He therefore traces the difficulty involved in the capture of first principles to the empirical side of philosophy. In our direct experience we are not immediately aware of any clear-cut analysis into the various elements which make it up, i.e. sense-objects, the structural features of space and time and other more general features. In direct experience the structural and other features to be found there are, as it were, blended into a whole. The analysis of experience forms, therefore, the starting point for thought, endeavouring as it does to analyse out these entities and relations from our perceptual field.

But what, it may be asked, has this account of immediate experience to do with the difficulty in the discovery of first principles? Well, simply this: Whitehead distinguishes between our observation of ordinary everyday objects and the way we become aware of the more general features of our experience. We habitually observe the things around us by what he calls the method of difference. For example, we are sometimes aware of an object, say an elephant, and we cannot help sitting up and taking notice, whilst in its absence we do not feel very perturbed. But this method of observation will not do when we come to consider first principles, for unlike our everyday objects, they are continually present in experience—so much so that our attention is usually not drawn to them.

We can now understand what Whitehead is getting at when he asserts: 'The metaphysical first principles can never fail of exemplification. We can never catch the actual world taking a holiday from their sway',[1] as they are always present in our experience (e.g. the logical framework of order) whether we take note of them or not. When it therefore comes to the discovery of metaphysics, we would not make much progress if we merely restricted ourselves to systematizing detailed matter of fact. This method (that of rigid empiricism) breaks down for the simple reason that, owing to the generality of the principles involved, we cannot be directly aware of them in the same way as we can, for example, of elephants and motor-cars. They can only be disengaged from the particular facts under observation as a result of a process of generalization. What we then need to make headway in our investigations is a generalized scheme, the elements of which should hang together, be logically consistent and applicable to all our experiences.

[1] P R, p. 5.

IV

We find that for such a construction to be successful it must, for Whitehead, satisfy the following conditions. It must in the first place originate from the generalization of certain features to be discovered in the various topics of human interest. He proceeds to give a lengthy list which, in marked contrast to his former restricted view, includes not only physics, but also physiology, psychology, aesthetics, ethical beliefs, sociology and language.[1]

It is obvious, therefore, that Whitehead is endeavouring to formulate a scheme which shall embrace every variety of experience; its general notions, derived from some particular set of facts, should be found capable of application to every type of fact. The success of the imaginative generalization (i.e. the scheme) is therefore to be tested by seeing whether it is applicable beyond the narrow range of experience from which it originated. For example, if originally derived from physics, it should not deal with notions which apply merely to that science, as we then get a variety of physicalism. It should also be applicable to other fields of experience —say, aesthetics. This may at first sight seem a very ambitious project, but Whitehead would no doubt answer that since both physics and aesthetics deal with aspects of human experience, they must satisfy the same conditions, which he believes are present in every kind of experience: for example, the respective sets of data they study are equally arranged in space and time. And even if there is a radical difference between the sorts of space and time they deal with, nevertheless Whitehead would say that they exhibit a common structure or are pervaded by a similar kind of order.

Whitehead further points out that in his description of philosophic method 'the term "philosophic generalization" has meant "the utilization of specific notions, applying to a restricted group of facts, for the divination of the generic notions which apply to all facts" '.[2] Now the above method of philosophic generalization bears some resemblance to Whitehead's attempt in S M W to apply the concept of the variable to the analysis of experience. By means of this notion, we can generalize particular features (entities and relations) in our experience, and thus arrive at abstract patterns of relationships which owing to their generality 'might hold for an indefinite variety of other occasions, involving other entities and other relations between them'.[3]

[1] P R, cf. p. 6. [2] P R, p. 6. [3] S M W, p. 32.

Whitehead's philosophy would seem then to link up (at least on its rational side) with the application of a mathematical or logical technique to nature. When speaking of this scheme he always seems to have at the back of his mind a super-deductive system applicable to the universe, a logic of nature whose abstract ideas reflect the interconnectedness of events.

Hence, although the method adopted by 'speculative philosophy' seems to make its first appearance in P R, we think that such a view was foreshadowed in S M W (as well as in his earlier writings) in his account of the function mathematics performs in thought. Apparently it attempts to go the whole way in giving us a complete analysis. It separates the empirical elements from the purely abstract conditions which they exemplify. In this case, by abstracting the empirical content from the structure of related events, we obtain an abstract pattern of relationships which forms the general conditions Whitehead is talking about.

We can now understand what he means when he tells us that the totality of these general abstract conditions is itself interconnected in the manner of a pattern with a key to it; so that from a select set of such conditions derived from any particular fact we can by reasoning develop a pattern involving an indefinite variety of such conditions. All that we then need to do in order to obtain such a scheme is to know something which is perfectly general about the elements in any experience. We are then able by deductive reasoning to proceed to know an indefinite number of equally general concepts. The resultant pattern or scheme of relationships Whitehead assumes to be an abstract representation of the relational system of events.

It is clear, then, that the function which Whitehead assumes mathematics performs in thought bears a strong resemblance to what he now calls philosophic generalization, whilst the totality of these general conditions (or patterns of relationship) comes very close to his notion of a philosophic scheme (at least its rational side) which deals with the most general features found in experience. It becomes apparent that the sort of generalities which mathematics in its broadest sense (i.e. symbolic logic) was looking for in S M W is very like that which philosophy seeks in P R.[1]

[1] Dewey (*Philosophical Review*, 1937, p. 174), sees in the definition of speculative philosophy in P R a suggestion that Whitehead would have us adopt a mathematical model and pattern in philosophizing. The task of a philosopher would thus be to discover in immediate experience the

It should be noted that this has nothing to do with the question whether Whitehead is right or wrong in his belief that mathematics investigates abstract conditions (or patterns of connection) derived from our experience of an order in nature. We are merely concerned to point out that there is some resemblance between (*a*) the method outlined here and the abstract conditions it deals with, and (*b*) that employed in P R and the general features making up the rational side of the philosophical scheme.

One further comment may be made: Whitehead's whole account depends upon the assumption that nature is a coherent structure of events. Only when this is granted are we justified in saying that we can disengage this structure at any particular point of fact, and then proceed by reasoning to any other section of it. That there is an abstract pattern of relationships may be true and interesting, but this is not by itself a guarantee that it exists in that form in nature. What he seems to be doing is to argue that because mathematical schemes apply so well to fact, the order they represent is also a characteristic of it. However, we have no justification for supposing that such systems necessarily stand in one-one correspondence to our experiences. Whether they do or not, as Whitehead himself recognizes, is largely a question for empirical investigation.

V

Whitehead continues his discussion by saying that: 'The second condition for the success of imaginative construction is unflinching pursuit of the two rationalistic ideals, coherence and logical perfection'.[1] By the latter notion he would at least seem to mean that the scheme must be logically self-consistent. He gives as an example of the importance of this notion the part played by mathematics in the field of natural science. 'The history of mathematics', Whitehead tells us, 'exhibits the generalization of special notions observed in particular instances. In any branches of mathematics, the notions presuppose each other. It is a remarkable characteristic of the history of thought that branches of mathematics developed

elements that can be stated in a succinct system of independent definitions and postulates, 'they being such that when they are deductively woven together there will result a coherent and necessary system in which "each element shall have the character of a particular instance of the general scheme" '.

[1] P R, p. 7.

under the pure imaginative impulse, thus controlled, finally receive their important application.'[1] But, as he points out, time may be needed. The theory of conic sections, for example, had to wait eighteen hundred years until Kepler in his laws of planetary motion gave it a practical application. 'In more recent years, the theory of probability, the theory of tensors, the theory of matrices are cases in point.'[2]

Whitehead therefore seems to hold the view that mathematics employs the method of imaginative generalization (the play of a free imagination controlled by the requirements of coherence and logic) that he has been urging philosophy to adopt. In mathematical thinking we find (i) the generalization of specific notions exemplified in particular instances (in this way we arrive at the postulates and definitions upon which the abstract system is based). We also find (ii) that in any particular branch of mathematics the notions presuppose or imply each other (i.e. these generalizations are imaginatively schematized), thereby giving us a deductive system. Finally the possibility always remains open that (iii) such abstract mathematical systems may receive an important (practical) application (they may be found applicable to certain sets of physical phenomena).

We may quote Huntington, who makes the same point as Whitehead when he tells us that the main interest of an abstract science centres in the logical relations between abstract propositions, rather than about the applicability of these to concrete things. 'But,' he goes on, 'many important mathematical theories have been developed as "abstract sciences", from an apparently quite arbitrary set of postulates, which have later proved to be powerful tools in applied mathematics, when important practical systems that satisfied all the postulates of these particular theories unexpectedly presented themselves.'[3]

Hence any postulate system, where we concern ourselves exclusively with the logical relations between abstract propositions, must also be classified under Whitehead's heading of an imaginatively generalized scheme. By its important application he would seem to refer to its interpretation by empirical data, or the discovery that there are practical systems which have the same structure as this abstract system. Hence when Whitehead in PR

[1] PR, p. 7. [2] PR, p. 7.

[3] In *Monographs on Modern Mathematics*, (Editor J. W. A. Young) p. 166.

talks of the important application of an imaginatively generalized scheme, he has as a model before his mind the interpretation of an abstract system by particular matters of fact. In this case the interpreted system is the cosmological (or philosophical) scheme, with its rational side—i.e. the logical framework—and its empirical side—the empirical relations and relata which satisfy the conditions laid down by this abstract scheme.

We may give a more precise specification of the abstract system elaborated in P R by identifying it with the extensive scheme (the basic postulates of which Whitehead gives on pp. 417-20) which becomes the cosmological scheme when a determination is given to it by the actualities of our present cosmic epoch. We can therefore consider this system from two distinct points of view: (*a*) from the point of view of its own internal properties as an abstract system (i.e. its rational side), where we deal with a coherent, logical and necessary system of general ideas and which is on the same plane as the abstract systems studied by logic and pure mathematics; and (*b*) from the point of view of its empirical exemplification—with a practical system which satisfies the conditions laid down by this abstract scheme.

Whitehead even in his earliest published work was interested in the application of symbolic logic to experience, for he told us: 'Ordinary algebra in its modern developments is studied as being a large body of propositions, interrelated by deductive reasoning, and based upon conventional definitions which are generalizations of fundamental conceptions. Thus a science is gradually being created which by reason of its fundamental character has relation to almost every event, phenomenal or intellectual, which can occur'.[1] The view put forward by Whitehead in the above passage certainly suggests his position in P R when he deals with the construction of an imaginative scheme. He believes that we come by such a scheme as the result of the generalization of certain basic features in experience, the derivative concepts being then interrelated by deductive reasoning (i.e. imaginatively schematized). Finally the system of general concepts thus arrived at is applied to the whole range of our experiences.

In the light of the above, one is not surprised to find that Whitehead was in U A also of the opinion that 'The ideal of mathematics should be to erect a calculus to facilitate reasoning in connection with every province of thought, or of external experience,

[1] U A, p. viii.

in which the succession of thoughts or of events can be definitely ascertained and precisely stated. So that all serious thought which is not philosophy, or intuitive reasoning, or imaginative literature, shall be mathematics developed by means of a calculus'.[1]

It should be observed that Whitehead is here using the notion of a mathematical calculus in the very general sense of a deductive system. He tells us: 'Mathematics in its widest signification is the development of all types of formal, necessary, deductive reasoning', that its sole concern 'is the inference of proposition from proposition'.[2] It is clear that Whitehead had as an ideal before his mind the construction of an abstract system by means of which we can definitely ascertain and state in a precise form the succession of thoughts and events. He thereby hopes to confer order upon our fragmentary experiences, and thus demonstrate that they presuppose one another, just as propositions do in a deductive system. We are thus enabled by the technique of symbolic logic to extend deductive reasoning to fields of enquiry which are usually not supposed to be amenable to mathematical treatment. At this point we need to indicate the similarity of the proposed calculus (i.e. 'a science . . . which by means of its fundamental character has relations almost to every event phenomenal or intellectual which can occur') with the endeavour of speculative philosophy to frame a system of general ideas in terms of which all our experiences can be related.

Though Whitehead at that time thought philosophy was not amenable to such treatment, this does not mean that he still maintains such a view in P R. It would appear from the context and the juxtaposition of philosophy with imaginative literature that his conception of philosophy when he wrote U A was very different from his later view. What he probably meant by 'philosophy' in 1898, may perhaps be seen from a work published twenty-two years later. 'By a metaphysical interpretation I mean any discussion of the how (beyond nature) and of the why (beyond nature) of thought and sense-awareness.'[3] The meaning he gives to 'metaphysics' here is just the reverse of his later position, which bears some resemblance to what he then termed the 'philosophy of science', where 'we seek the general notions which apply to nature'.[4] In other words, 'Our sole task is to exhibit in one system the characters and inter-relations of all that is observed'.[5] Hence, Whitehead's earliest

[1] U A, p. viii. [2] U A, p. vi. [3] C N, p. 28.
[4] C N, p. 28. [5] C N, p. 185.

pronouncement in no way conflicts with his 'philosophy of science' —his later view in P R and his assertion that when symbolic logic has sufficiently expanded it will become the foundation of aesthetics, as well as of ethics and theology.[1] Indeed, this programme is more radical than the role assigned to the proposed calculus in U A.

VI

Whitehead asserts that the chief error in philosophy is what he calls overstatement.[2] That is to say, he thinks philosophy is working on the right lines when it tries to generalize the elements in our experience, but the success of this method can be exaggerated. This gives rise to two forms of overstatement. The first is 'the fallacy of misplaced concreteness', which consists in considering experience in terms of such notions as space, time, matter and mind, and then by supposing them to have an independent existence gives them a concreteness they do not possess. It also causes us to neglect the rest of our concrete experience from which these notions have been abstracted.

An example of a philosophic scheme which falls into such a fallacy is the world pictured by the materialist, made up of bodies moving rapidly in space and time. Another offender is the substance-quality philosophy of Locke, which also gives us an over-simplified edition of fact very different from the patterned whole we are aware of in perception. Whitehead, no doubt wishing to impress upon us the inadequacies of such schemes which fail to deal with experience as it really is, claims that we can measure the success of a philosophy by its comparative avoidance of this fallacy.[3] He obviously takes agreement with immediate experience as the standard by which any philosophic scheme must be tested. Whitehead's philosophy does not therefore attempt to explain the universe or to show how it functions, but merely sets out to define the most general features of things.

The other form of overstatement 'consists in a false estimate of logical procedure in respect to certainty, and in respect to premises'. What Whitehead means by this can best be seen from the following: 'Philosophy has been haunted by the unfortunate notion that its method is dogmatically to indicate premises which are severally

[1] *Philosophical Review*, 1937, p. 186. P R, cf. p. 9.[2]
[3] P R, cf. p. 10.

clear, distinct, and certain; and to erect upon those premises a deductive system of thought'.[1] At a first glance, it may seem that Whitehead is attacking the procedure we have suggested he uses in his method of imaginative generalization. And if this is the case, it certainly appears to contradict our interpretation of it. But the view we have outlined above is very different from the one criticized here. It gives a radically different status to the premises from which the system starts. For Whitehead they are in no way self-evident, but have a purely hypothetical character. They are generalizations (or abstract concepts) based upon certain features discriminated by us in experience which are afterwards elaborated into such a system.

Further, as we have seen, the general principles delineated in the resultant scheme should also be exemplified in other facts of experience to which it is applied. This universality of application remains, however, an ideal, since we have no guarantee, logical or otherwise, that the premises upon which the scheme is based are metaphysically necessary, i.e. that they will apply to all events. As Whitehead points out, owing to the limitation of our powers of observation our metaphysical notions have only an approximate character. They are really ideal limits to such ultimate principles. Hence, when we apply such an abstract scheme to experience, even though the process of inference is perfectly rigorous, the conclusions reached are no more certain than the premises from which we start. Whitehead thinks it doubtful whether we can even attribute metaphysical necessity to the most general principles of logic and mathematics, having regard to the mistakes in the past respecting Euclidean Geometry.

Arithmetic, considered as an abstract system, would seem to be perfectly valid and to hold independently of any empirical facts, but when we come to give it a practical application, then the possibility is always open that in some other cosmic epoch, e.g. in a fluid or gaseous world, arithmetic might find no useful application. In metaphysics, unlike mathematics, our chief interest lies not in the self-consistency of the abstract scheme, but in giving a concrete interpretation to it. We are interested in finding some actual system in which the scheme is exemplified, or, rather, to be more precise, some abstract system whose structure is identical with that of our metaphysical situation (or cosmic epoch). The search largely resolves itself into a question of practice; we endeavour by a

[1] PR, p. 10.

process of trial and error to find some general scheme which will approximate to this ideal.

Whitehead therefore points out that, unlike mathematics, 'The verification of a rationalistic scheme is to be sought in its general success, and not in the peculiar certainty, or initial clarity, of its first principles'.[1] Such a scheme is to be verified in the same way as any hypothesis—namely, by its general success in practice—and does not depend for its truth upon the peculiar certainty or initial clarity of its premises, as Descartes thought. Whitehead argues that under the influence of mathematics philosophy has erroneously accepted deduction as its standard method, instead of taking it 'as an essential auxiliary mode of verification whereby to test the scope of generalities'.[2]

Whitehead believes that 'the primary method of philosophy is descriptive generalization'.[3] By this he would seem to mean that the premises upon which the philosophic scheme is based must have an empirical foundation; they should be derivative from experience by a process of generalization (or abstraction). But because of this it ought not to be assumed that Whitehead rejects the use of the deductive method in philosophy. Nothing could be further from the truth, as once we have arrived at our general principles it plays an important secondary role—schematizing them into a coherent system. We are thus enabled to test their applicability beyond the limited field from which they originated. As he puts it: 'empirically the development of self-justifying thoughts has been achieved by the complex process of generalizing from particular topics, of imaginatively schematizing the generalizations, and finally by renewed comparison of the imagined scheme with the direct experience to which it should apply'.[4] We note once again that three distinct processes are involved in Whitehead's philosophic method—(i) the abstraction of the basic principles from particular matters of fact, (ii) the process of deductive inference by which we make clear the implication of these premises, and (iii) the application of the derivative scheme to the facts of experience.

If we must look for comparisons, Whitehead's approach comes much closer to that of the mathematical physicist than to that of the pure mathematician. There is this important difference in their respective attitudes: one can be said to treat mathematics as a tool in its application to natural phenomena, the other as an end in itself. The mathematician depends for the validity of his system

[1] PR, p. 10. [2] PR, p. 14. [3] PR, p. 14. [4] PR, p. 22.

upon rigid proof and logical consistency. The mathematical physicist, however, does not merely rely upon logical criteria for the truth of his scheme. He appeals to the facts themselves, checking his approximate, mathematically expressed theory directly with experience, discarding the theory if it is found to conflict with fact.

Though a philosophical scheme, like any other deductive theory, needs to satisfy the demands of logical coherence, this is not, as Whitehead admits, the only criterion of its truth; it must also agree with experience. But he seems to take the further step and assumes that experience is coherent too, i.e. the elements presuppose each other. He argues that it is precisely this fact which makes the scheme capable of application. However, it is evident that this contention is already an integral part of the material content of his theory, as distinct from the coherence required by its formal structure. It is one thing to assume that a theory must be coherent in the latter sense and quite another to assume that experience has this characteristic. Formal coherence solely depends on logical criteria, whilst the belief that experience also is coherent depends upon our direct observation of the facts. Such a scheme, when it obtains a practical application, can at the most be said to give us an approximate account of experience, i.e. act as a model for it. Whitehead, if pressed, would undoubtedly agree to this. For him our metaphysical principles as yet formulated have only an approximate character, though he does not rule out the possibility that if our intelligences were more perfect we might acquire precise knowledge of these principles.

VII

Whitehead asserts that: 'Whatever is found in "practice" must lie within the scope of the metaphysical description'.[1] By this he seems to mean that when we put our theory into practice we should find it to be applicable to every variety of fact. If it is not, the scheme must be revised in such a fashion that it can be thus applied. Nevertheless he warns us: 'No metaphysical system can hope entirely to satisfy these pragmatic tests. At the best such a system will remain only an approximation to the general truths which are sought'.[2] Hence it follows that there are no precisely stated axiomatic certainties from which we can start in philosophy. Our approach has rather to be

[1] P R, p. 17. [2] P R, p. 17.

the other way round—from tentatively formulated principles we should proceed to ones more precisely stated.

The proper objective of philosophy would, then, seem to be 'the gradual elaboration of categorial schemes, definitely stated at each stage of progress',[1] in other words, to elaborate such general schemes in the same way as science puts forward hypotheses, and continues in practice in the light of fresh evidence either to amend them or to introduce new ones in their stead. Now this procedure resembles the method of approximation used, e.g. in mathematics, when we wish to arrive at the solution of some complicated problem. We first posit an abstract solution, and then gradually by tests and experiments bring it into harmony with the actual situation. The Socratic method (dialectic) comes in principle to much the same thing, for we first select a definition at random (of, say, justice) and then in the ensuing discussion bring it nearer to the ideal definition. In the same sort of way, Whitehead thinks we should state our metaphysical principles tentatively as hypotheses and then by progressive modifications approximate to the ideal scheme.

This view that the principles of our philosophic schemes are really approximations is related to Whitehead's assertion in A I that 'speculative philosophy embodies the method of the "working hypothesis" '.[2] As is well known, such a method is of use in science as it collects and brings together a large number of apparently diverse facts under one point of view (or scheme). The resultant scheme serves as a guide to our investigations by keeping before our minds the precise phenomena which need an explanation—and leads to the discovery of the general law they follow. In the same manner Whitehead seems to assume that the use of a working hypothesis in philosophy will enable us to discover the basic principles underlying the universe—or at least go some way towards it.

In one important respect, however, the scheme put forward by Whitehead differs from the working hypothesis used in most fields of enquiry, as it has some of the characteristics of a deductive system. For example, there is no reason to suppose that either the 'Theory of Evolution' or the 'Materialistic Conception of History', as at present stated, is a coherent, logical and necessary system. Though of course, there is no reason why they should not be thus stated. The working hypotheses generally used by us lack the logical rigour and precision of the deductive system. So though

[1] P R, p. 10. [2] A I, p. 286.

deductive systems are working hypotheses, it is clear that not all working hypotheses are deductive systems. Most of the theories employed by such sciences as biology, psychology and sociology are not strictly formal. Indeed, it is just this absence of deductive form which prevents our reasoning in these disciplines with the same precision as in the mathematical sciences.

Now our contention that deductive systems have a hypothetical character may perhaps be questioned. It may be argued that, if there is one field in which we can arrive at necessary truths, then surely it is in the realm of logic and pure mathematics. On this point we may again refer to Huntington, who brings out the difference between the older mathematical method which one finds illustrated in Euclid, and the more modern method of postulates used nowadays by pure mathematics and symbolic logic.

He points out that, according to the older habits of thought, modelled after a demonstration in Euclid, and which had for its goal the attainment of ultimate truth, the only way to demonstrate a proposition in science or philosophy was to start with obvious axioms and then proceed by successive logical steps until the desired proposition was proved as a theorem. Huntington continues, 'The postulational method on the other hand sets itself a much more modest aim. It makes no use of self-evident axioms. It does, indeed proceed from hypothesis to conclusion by processes of logic, but it makes no claim that its hypotheses are self-evident or that its conclusions are true. All that it claims is that if its hypotheses are valid then the conclusions will also be valid. The postulational method is the "If-Then" method; its results are never absolute, always hypothetical'.[1]

It will be seen that Whitehead's philosophic method differs from that used by Descartes and Spinoza, in much the same way as the method of postulates differs from the method of axioms. Hence, when Whitehead criticizes past philosophy for following in the footsteps of mathematics (i.e. dogmatically indicating premises which are clear, distinct, and certain, and erecting upon these premises a deductive system of thought) he is primarily criticizing the older habits of thought modelled after a demonstration in Euclid. The most famous example of this tendency in philosophy is Spinoza's *Ethics*, where we have a rigid hierarchy of propositions, starting with obvious axioms and demonstrated theorems. In this way Spinoza hoped to give philosophical propositions the same

[1] *Philosophy of Science*, 1937, p. 485.

universality and necessity as was claimed for mathematical propositions.

Whitehead's philosophy, on the other hand, is modelled rather on the method of postulates. His premises are in no way self-evident, but have a purely hypothetical character. Neither are the conclusions reached in any way more certain than the premises on which they are based. The resultant metaphysical scheme has therefore only a tentative character, and depends for its truth upon its empirical verification, and not merely upon logical criteria. It must be frankly admitted that large parts of Whitehead's philosophy have not the logical rigour of a deductive system. Such a system would seem to be rather a logical ideal to which philosophy should aspire.

We might perhaps be tempted to say with Russell that philosophy deals with the sciences only when they are vague and unspecialized, that philosophy is science at its undeveloped levels forming a matrix from which the sciences are broken off one by one, as a result of their increasing specialization. Whitehead, however, sees the relationship between science and philosophy in a rather different light. Though the sciences look for general principles, they are limited by the boundaries of their subject matter. Philosophy, on the other hand, attempts to deal with general principles not restricted to any special topic. For Whitehead the main characteristic of philosophy is its complete generality like $\phi\hat{x}$—the special sciences being, as it were, the particular values, i.e. the philosophical scheme is the genus of which the particular sciences are the species. We must then distinguish between generality and vagueness. There is nothing vague about $\phi\hat{x}$ although it is certainly general. So with Whitehead's philosophic scheme: its principles, while general, cannot be termed vague, though we do not deny that his mode of expressing them may at times be disconcertingly obscure.

SUMMARY

The method of philosophical construction put forward by Whitehead resembles the hypothetico-deductive method used by the scientist and mathematician. Whitehead lays down certain criteria that a philosophic scheme needs to satisfy, namely, it must be coherent, logically necessary and applicable.

The function Whitehead assumes mathematics performs in

thought bears a strong resemblance to what he now calls philosophic generalization. The validity of his account depends upon his assumption that nature is a coherent structure of events in which the two principles of logical order and temporal process occur. However, there is no justification for assuming that logical systems have a one-one correspondence to experience.

Whitehead's approach comes closer to that of the mathematical physicist, where mathematics is treated as a tool in its application to natural phenomena, rather than as an end in itself. Philosophic schemes, he tells us, must not only satisfy the demands of logical coherency—they must also agree with experience. Philosophical principles are therefore regarded by him not as eternal truths, but rather as approximations—as working hypotheses.

CHAPTER II

WHITEHEAD AND LANGUAGE

AS language has such a marked interest for philosophers nowadays, it is worth noting the manner in which Whitehead's views on this topic diverge from the current approach with its emphasis on ordinary usage. Whitehead believes that in the sort of philosophy in which he is interested (which has for its task the bringing together of the world of psychological experience and physical nature into one system) the meaning of common words and phrases has to be stretched and given a wider generality.

Whitehead's terminology is undoubtedly at times puzzling. One may trace this to his desire to coin a language which should apply to any type of event—physical or psychological. In doing this, he borrows many terms from ordinary language, such as 'feeling', 'concept', 'mind', and 'God', assigning to them radically different meanings from those they have in ordinary use. One has therefore to exercise a certain caution to prevent oneself from giving such words their everyday meaning. Indeed, as they appear in the text of PR they are often used almost in a metaphorical way. When, for example, Whitehead talks of 'conceptual feelings', he usually has in mind the notion of logical structure, whilst by 'physical feelings' he generally refers to the transmission of physical energy in nature. If these phrases are interpreted psychologically, the meaning of the concepts expressed take on a much too specialized subjective character. We may be led to assume that Whitehead is talking about psychological states, when he is really referring to processes in the physical world. And this is the easy fallacy into which many interpreters of Whitehead have fallen.

Whitehead's views on language are closely tied up with his views on perception. Unlike the modern philosophical analysts with their appeal to everyday experience, he believes that the world of perceived objects (of substances qualified by predicates) only gives us an abstract picture of the physical world. The subject-predicate form of expression, its linguistic counterpart, is considered by Whitehead to be a high abstraction coined to deal with such situations. Language, he points out (and he is presumably thinking of an Indo-European language here), was designed to express such clear-cut concepts as 'green leaf' and 'round ball'.[1] Nevertheless, as he readily admits, the abstract system of concepts embedded in ordinary language has proved itself to be of great pragmatic value in enabling us to come to grips with our common-sense world. From his earliest writings onwards Whitehead crusaded against what he termed 'the fallacy of simple location'—the conception of the world as made up of independent objects (or substances) characterized by qualities. He regards nature as a system of events rather than a mere collection of static objects, and as having the character of passage about it. Whitehead also contends that the structure of this system may be more accurately expressed by multi-termed relationships than by subject-predicate ones. Not only do we become aware that nature has this dynamic character through the information presented by physical science, but also, he would claim, by our awareness of the causal world around us *via* our perceptions in the mode of causal efficacy.

For Whitehead, then, as with Bergson, ordinary language merely gives us a useful abstract for the purposes of life, and cannot therefore by its very nature grasp the dynamic character of events. He traces the tendency to rely on ordinary language as a criterion for fact back to the Greeks, pointing out that excessive trust in common forms of language vitiated the philosophy and physics of the Greeks as well as that of the Middle Ages. To bring this point out further, he quotes Mill's assertion that the Greeks believed that the classes of objects made for them by the popular phrases of their country were natural and all others arbitrary and artificial, and that by determining the notions attached to common language they would become acquainted with fact.[2] One cannot help feeling that Mill's criticism would also be applicable to certain types of modern philosophy which contend that ordinary usage gives us a more accurate picture of things than a description worded in technical terms.

With Whitehead's distrust of ordinary language as a guide to the true nature of 'reality' there goes a belief in the value of poetic language. He thinks that poetic language, by its stress on the aesthetic and dynamic side of our experience, will give us some measure of insight into the real efficacious world of events. But this,

[1] P R, cf. p. 234.

[2] P R, cf. pp. 15-16 (J. S. Mill. *Logic*, Book V, Chapter III).

as he is acutely aware, is not confined to poetry. Certain symbolic devices such as the burning of incense and religious ceremonials have often a greater efficacy in eliciting aesthetic experiences (which, he thinks, are closely tied to the efficacy of nature) than have written or spoken words.

There may appear to be some inconsistency between Whitehead's emphasis on the efficacy of poetic and literary language in giving us an insight into the dynamic and aesthetic aspects of nature, and his stress on the adequacy of mathematical formulae for describing the physical world.[1] However, the conflict is only apparent. He would argue that whereas mathematical description largely concentrates on the structure of events, poetic language emphasizes the dynamic aspects of our experience.

II

Whitehead believes that sentences are ambiguous as to the exact propositions (or meanings) they indicate. The reason for this is not far to seek. He considers a proposition to have a specific reference to psychological elements—to thoughts and images—as well as to the context of utterance. The meaning of a sentence will thus vary in accordance with its psychological and environmental contexts. He takes as an example, the sentence 'Socrates is mortal'. It may mean, we are told, 'The man Socrates is mortal' or 'The philosopher Socrates is mortal'. In the latter case, we usually know more about the connotation of Socrates—roughly the sort of things he did and said and the circles he moved in—than in the first case. Or consider another example, 'Caesar crossed the Rubicon'. He argues that such a sentence has no one simple meaning, since it symbolizes an indefinite number of diverse propositions. It might refer to a Roman soldier actually witnessing Caesar's crossing, or a modern Latinist visiting the river today, and going over in his mind the past events. In the first case there would be a reference to the actual happenings occurring before the observer's eyes, in the second to particular thoughts and images in the observer's mind, as well as to the present environmental context.[2]

Whitehead thus differs from those moderns who tend to identify the sentence with the proposition, and who also believe that one

[1] As he puts it, algebra, unlike ordinary language, endeavours to exemplify in its written structures the pattern which it is its purpose to convey.

[2] P R, cf. pp. 276-279.

can study language as a pure science apart from any direct reference to thought and things. For example, the modern semanticist seems to assume that, by the device of quotation marks, language can be studied as if it were another type of object. As a result of his emphasis on context, Whitehead regards language as being highly elliptical. He illustrates this by pointing out that, according to the context of utterance, the word 'Caesar' may mean a puppy dog, a negro slave or the first Roman Emperor. Whitehead, however, though he stresses context, would not, like some philosophical analysts, take it as a complete substitute for any psychological reference. In his approach to language, then, Whitehead emphasizes the need for interpreting language in its psychological as well as environmental setting. Most modern work on communication, for example, is not interested in the psychological processes underlying the communicated symbols. If the linguistic symbols are the same, the two processes are usually regarded as similar. However, we cannot always deduce from a simple examination of the linguistic elements involved the precise meaning which the speaker or writer intends to convey.

SUMMARY

Whitehead's views on language diverge from the current approach with its emphasis on ordinary usage. His position is closely linked with his theory of perception. The philosophical account of perception in terms of substances qualified by predicates only gives us, he argues, an abstract picture of the real world. Whitehead distrusts ordinary language as a guide to the true nature of experience, as it fails to grasp the dynamic character of events. He argues that poetic language, with its stress on the aesthetic and dynamic side of our experience, gives us a deeper insight into reality. But such an account, he would go on, needs to be supplemented by logico-mathematical description, which makes clear to us the structure of things.

CHAPTER III

NATURAL THEOLOGY

ONE of the difficulties philosophers have had in understanding Whitehead's Natural Theology has been due to their failure to grasp precisely what he means when he speaks about Deity. We believe that the only way this concept can be understood is (i) to view the so-called theological doctrines of Whitehead against the background of his earlier work, and (ii) to attempt to translate some of the more obscure passages of his later work into the comparatively lucid terminology of his earlier writings. This we propose to do.

In S M W Whitehead spoke about the 'realm of eternal objects', or the 'general system of the relatedness of all possibilities'.[1] This concept has a certain similarity to what in P R he calls the 'extensive continuum', which he conceives of as 'one relational complex in which all potential objectifications find their niche'.[2] It underlies, he tells us, the whole world, past, present, and future. For Whitehead it has the purely formal character of an abstract scheme which can be given any kind of perceptual or physical content. At the same time he indicates that it lays down certain logical conditions (of order) to which future events must conform.

This description of the extensive continuum fits in with Whitehead's account of the Platonic Receptacle in A I, and is an example of the way in which the same concept takes on a different guise in different works. By the 'Receptacle' he means the 'general interconnectedness of things' (or extensive scheme) which unites the atomic events into one common world. He stresses its formal character by telling us it 'may be conceived as the necessary community within which the course of history is set, in abstraction from all the particular historical facts'.[3] Whitehead thus regards the scheme as a bare logical form in which a particular type of content (i.e. events) may be fitted, just as one can assign a value (or give a determination) to a propositional function. In this connection it is interesting to note Russell's description of a propositional function

[1] Cf. the chapter on 'Abstraction'. [2] P R, p. 91. [3] A I, p. 192.

as 'a mere schema, a mere shell, an empty receptacle for meaning, not something already significant'.[1] One thing seems clear. The extensive continuum, the Receptacle, and the realm of eternal objects, all seem to refer to the same general system of undetermined relationships. We deal with patterns of connection in abstraction from particular events.

Now it may be asked, what has all this to do with Whitehead's conception of God? As we shall see, the system of general relationships is identical with the 'Primordial Nature of God', having all its logical properties. The apparent divergence between them arises from the fact that Whitehead describes the latter in terms which are much too concrete (making us give it an experiential content), since he is really talking about structure in its mathematical sense (i.e. sets of relations). When, for example, Whitehead speaks of the 'Primordial Nature of God', he describes it as the 'unconditioned actuality of conceptual feeling at the base of things'.[2] This language masks the abstractness of his account; even the term 'conceptual' has usually an experiential significance. As we shall later observe, 'God's conceptual feelings are empty of all experiential content', and merely refer to the abstract logical patterns within this scheme. When he therefore states that God is the actuality of conceptual feeling at the base of things, we need to recognize that he is merely referring to the n-adic ordering relation R, the logical framework in which events are related.

Whitehead thinks the 'Primordial Nature of God' to be the ordering entity in nature, since it relates the multiplicity of events into a well-ordered system. However, he tells us elsewhere precisely what he means by the order of nature in terms free from any theological construction. 'The order of nature, prevalent in the cosmic epoch in question, exhibits itself as a morphological scheme involving eternal objects of the objective species',[3] expressing the theory of extension in its most general form. In this theory the related events with their particular characteristics have been entirely eliminated, so that we are left with a purely formal structure of serial relations. It is this which provides the basis of the order in nature. There seems little doubt that 'the morphological scheme involving eternal objects of the objective species' is the same concept as the 'Primordial Nature of God'. He describes these eternal objects as 'the mathematical platonic forms', and further

[1] *Introduction to Mathematical Philosophy*, p. 157.
[2] P R, p. 486. [3] P R, p. 414.

tells us, 'eternal objects, as in God's primordial nature, constitute the Platonic world of ideas'.[1] The similarity between these two concepts can readily be seen if we look at what Whitehead thought the Platonic world of ideas to be. He took it as a 'refined, revised form of the Pythagorean doctrine that number lies at the base of the real world',[2] and he goes on to point out that, since the Greeks represented numbers by patterns of dots, their notions of number and geometrical configuration were less separated than are ours. Whitehead then in this context seems to be using the phrase 'Platonic idea' to refer to the concept of an extensive pattern which has some affinity to the notion of a geometrical configuration, since it is also a set of terms ordered by a serial relation. When Whitehead speaks about 'eternal objects' as in God's primordial nature, he is really referring to a system of such abstract structures.

II

Let us now examine sample passages illustrating the way Whitehead uses the notion of God in his philosophy, translating them as far as possible into logical language. If such a translation can be made, if for vague passages we can substitute more precise statements expressed in logically neutral terms, and if the resultant account fits in with the main body of his thought, we have then some reason for supposing that the 'Natural Theology' of Whitehead arises more from a change in terminology than from one in philosophical viewpoint.

In discussing the nature of God in RM Whitehead states: 'God, who is the ground antecedent to transition, must include all possibilities of physical value conceptually, thereby holding the ideal forms apart in equal, conceptual realization of knowledge. Thus, as concepts, they are grasped together in the synthesis of omniscience'.[3]

Expressing this passage in logical terms, we may say that the extensive scheme can be thought of as the ground antecedent to the creative advance of nature or transition, since it determines the logical conditions to which all events have to conform. It includes 'all possibilities of physical value conceptually', as this scheme (which is perfectly general and devoid of qualitative content) defines an infinite range of values. This range consists of all the specific perceptual or physical characters which can be obtained

[1] PR, p. 63. [2] SMW, p. 36. [3] RM, p. 153.

by giving every possible determination to the variables in $R(x_1 \ldots x_2 \ldots x_n)$. It may be said to hold these abstract patterns (logical forms) apart, since they stand in relationships to each other within this abstract system (or conceptual realization of knowledge). Whitehead's use of the term 'knowledge' in this context may seem a little strange, but by it he merely seems to mean such an infinite range of possible values. Only in this way are these bare logical forms (or concepts) related or grasped together in the systematic complex (or synthesis of omniscience). It is evident, however, that Whitehead's description in theological terms is much too concrete, and cloaks the abstract logical character of his whole account.

When he says that God (or this scheme) is complete since 'his vision determines every possibility of value',[1] he means, as we have already indicated, that the extensive scheme is perfectly general, and can have any type of qualitative content given to it. Thus the events we are aware of in our everyday life may differ from each other as much as a tea party differs from an atomic explosion. But what they must have in common, is the same type of logical structure; they must be related by means of this logical framework into one common world. By 'God's vision', Whitehead is here referring to the systematic complex of extensive relationships. As this scheme determines such an infinite range of possibilities, Whitehead asserts 'his knowledge of the relationships of particular modes of value is not added to, or disturbed, by the realization in the actual world of what is already conceptually realized in his ideal world',[2] i.e. by the particular realization of any member of the class of values defined by it. If we neglect the difference in terminology, the notion of God as used in this context comes to very much the same thing as the extensive scheme.

Whitehead believes that, apart from these logical conditions (or framework of extensive order) which on his account determine the actual structure exhibited by events, there would be no guarantee that future events will be ordered in the same way as the events in the present. From this point of view God (or the scheme) may be considered as the principle of limitation or concretion, since it limits the course of events to being ordered in this manner. It therefore lays down conditions to which all events have to conform. It is this, he thinks, which makes induction possible. If our predictions are to be in any way justified, then underlying all our

[1] R M, p. 153. [2] R M, p. 154.

arguments must be the presupposition that the order of nature will continue in the future.

Whitehead in P R repeats what he has already said in a somewhat different manner. 'God', he tells us, 'is dipolar. He has a primordial nature and a consequent nature. The consequent nature of God is conscious; . . . The primordial nature is conceptual, the consequent nature is the weaving of God's physical feelings upon his primordial concepts'.[1]

We have already touched upon many of the points contained in this passage. It illustrates how Whitehead, in some of his more metaphysical passages, ceaselessly repeats the same ideas, throwing them in different combinations. He is telling us that, as is the case with all actualities, the scheme can be considered from two aspects, (i) its logical aspect, as an ordering relation *R*, entirely unlimited in its range of possibilities; and (ii) its physical aspect, as relating the multiplicity of events into one common world. In (ii) the scheme has a definite physical or qualitative content given to it by the course of events; i.e. the logical relations are interwoven with the individual peculiarities of events. It is difficult to follow Whitehead when he says that the 'consequent nature of God is conscious', but it is questionable whether he means by this more than that we are consciously aware of the concrete world in perception.

At this stage the reader may be feeling rather puzzled. He may ask; if Whitehead in his account of God is really only dealing with certain logical features of our experience, why has it been assumed that he was erecting a 'Natural Theology'? Why has it been taken for granted that he is simply talking about moral and aesthetic values? It seems evident, especially if one remembers Whitehead's early training, that the values he is dealing with are more of the sort one meets with in books on mathematics and mathematical logic, than in those found in works on ethics, aesthetics and theology.

The difficulty, of course, lies in the language used, as the words of everyday life do not adequately express the abstract notions permeating Whitehead's philosophy. Russell's criticism of the cheap popularizations of modern physics applies equally well to Whitehead's deification of the extensive scheme. Russell points out that our ordinary everyday language is totally unsuited for expressing the abstract concepts of physics, and that only mathematics and mathematical logic can say as little as the physicist means to say.

[1] P R, p. 488.

'As soon as he translates his symbols into words, he inevitably says something much too concrete, and gives his readers a cheerful impression of something imaginable and intelligible, which is much more pleasant and everyday than what he is trying to convey.'[1]

III

It is worth while examining some of the antitheses which Whitehead formulates at the end of PR, since they state in a succinct form his 'Natural Theology'. It will also enable us to see how well our interpretation fits his account. The origin of these antitheses can be traced to the method adopted in his discussion. Whitehead's usual procedure is to consider each such antithesis from two different points of view, the first part from a logical standpoint and the second from a physical or experiential. It is this deliberate change of attitude which generates the apparent self-contradictions. Consider the following examples:

I. 'It is as true to say that God is permanent and the World fluent, as that the World is permanent and God is fluent.'[2]

Now since by 'God' here Whitehead is referring to the extensive scheme, it must as an abstract pattern of logical relationships be considered permanent (or unchanging). On the other hand, the actual world or course of events is fluent, since it has the character of temporal passage about it. But if we shift our attention to concrete experience, the actual world as pervaded by this 'framework of order', is to be thought of as permanent. The scheme relates the multiplicity of events into one system or common world made up of many enduring objects. On Whitehead's view, the physical objects we meet with in everyday life are really routes of events ordered by a serial relation into such enduring systems having the character of permanence about them. However, in our immediate field of experience, the scheme of extensive order may be conceived of as fluent, since we are aware of it in the form of a related system of finite events, illustrated by a many-hued variety of colours, sounds, and other sense-qualities.

Hence, when Whitehead elsewhere asserts that the abstract scheme embodies 'the principle of unbounded permanence acquiring novelty through flux',[3] he can be said to mean that, with the creative advance of nature, the extensive scheme stretching into the as yet unrealized future is actualized into definite novel

[1] *Scientific Outlook*, p. 85. [2] PR, p. 492. [3] PR, p. 234.

events. On the other hand, when he says that the course of events embodies 'the principle of bounded flux acquiring "everlasting permanence" ', he would seem to mean that the logical scheme transforms the multiplicity into a uniform world-system made up of enduring objects, planets, trees and stones, which are sub-societies in this larger society or universe. For Whitehead, then, the physical things which endure are societies having a historicity about them. An event, for example our present observation of the table, is merely a cross-section of such a dynamic history.

II. 'It is as true to say that God is one and the World many, as that the World is one and God many.'[1]

We have already touched on this antithesis; the extensive scheme taken as a logical system is a unity (or a continuum) whilst the actual world made up of atomic events can be considered as having a multiform character. But looking at the latter concretely, we experience it as a unity, since the events are related by their extensive relations into one system. Conversely, the extensive scheme viewed as pervading the actual world can be considered a multiplicity. With the creative advance of nature the scheme is atomized into definite events (or routes of events) spreading themselves spatially and temporally, and tied together by serial relations into enduring objects. In more mundane language, we perceive it as made up of objects related together by spatio-temporal relations.

III. 'It is as true to say that God transcends the World, as that the World transcends God.'[2]

What Whitehead seems to be saying here is that the extensive scheme, considered in its full generality, transcends the actual course of events. Or, as he puts it, 'these general abstract conditions transcend any particular course of events'. They are more fundamental than the four-dimensional space-time continuum, constituting the present epoch, which is only one particular type of determination of the possibilities inherent in the logical pattern of relationships. In other epochs, as he points out, space might well have 333 dimensions. Further, the scheme in relation to the present world transcends it in the sense that it stretches into the future. It thereby determines conditions to which future events must conform.

However, if we view the scheme concretely, as relating the events in our experience, we refer to the 'framework of order'

[1] P R, p. 492. [2] P R, p. 492.

permeating fact. In this fashion, it can be said that the course of events (or the world) transcends the extensive scheme, since the latter though logically prior is (physically) derivative from the mutual relations between events. Whitehead is evidently thinking of its derivative nature when he refers to the scheme as the 'first determination of order' and to the 'Primordial Nature of God' as the 'primordial created fact'.

IV. 'It is as true to say that God creates the World, as that the World creates God.'[1]

If one considers the scheme purely logically, as a relational system underlying the world past, present and future, one might say that it is creative of the world, since its extensive relations determine conditions to which all future events must conform, imposing a common order on all that happens. Each extensive perspective, when actualized by the creative advance, therefore forms the initial phase in the development of a novel event. In other words, the extensive scheme enters into each developing event, giving it a certain logical structure or order. Hence, the extensive relations can be said to govern the event's development, as they determine the logical conditions to which it has to conform. In this way God may be said to be creative of the world.

On the other hand, viewing the extensive scheme as pervading the actual world, we may say that the world creates God, since this scheme is derivative from the relationships between events. This throws some light on what Whitehead means when he speaks of the 'Primordial Nature of God' as being a 'creature of creativity', though it limits the course of events to having that form of extensive order. From a physical or experiental standpoint it may be regarded as a 'creature of creativity', since it is derivative from the relationships between events in the actual world.

Whitehead completes his account by making what at first sight would appear to be a remark savouring of religious mysticism. We are told that, 'The theme of Cosmology, which is the basis of all religions, is the story of the dynamic effort of the World passing into everlasting unity, and of the static majesty of God's vision, accomplishing its purpose of completion by absorption of the World's multiplicity of effort'.[2] However, viewed against the background of the above discussion, this passage falls into its proper perspective. What Whitehead is really saying is that the theme of cosmology is the story of how the creative activity of the

[1] P R, p. 492. [2] P R, p. 494.

world (or course of events) becomes related within the scheme to give rise to our common-sense world, and of how the 'framework of order' is thus completed, being given a qualitative content by the course of events. When he therefore states that this process is the basis of all religion, the most he could possibly mean is that these permanent aspects of nature have been intuitively singled out from experience, and taken as a basis for particular religious systems.[1]

IV

Though our account has been formulated in perceptual terms, it applies equally well if translated into physical language. The reason for this is simple. As we shall see, Whitehead appears to have taken as a model for his description of the perceptual field the electro-magnetic field of physics.

Looked at from this standpoint, 'Creativity' becomes the field of force (or flux of energy)—the transmission of energy from event to event in physical nature. The 'Consequent Nature of God' or the 'physical prehension by God of the actualities of the evolving universe',[2] is in this context to be thought of as made up of events pervaded by electro-magnetic characters (i.e. lines of force) and related into one spatio-temporal world or physical field by means of this framework of order. Whitehead in PR definitely states that our present epoch is formed by an 'electro-magnetic society', which 'exhibits the physical electro-magnetic field'.[3] Further, he is at pains to emphasize that our everyday enduring objects are specialized sub-societies within the more general electro-magnetic society or spatio-temporal universe.

[1] In his paper 'Immortality' (ed. P. A. Schilpp, *Philosophy of A. N. Whitehead*), the interplay of God and the World becomes the two sides of the universe: the 'World of Value' and the 'World of Activity', and he identifies the former notion with the concept of God, cf. p. 694. He further tells us, 'The main thesis of this lecture is that we naturally simplify the complexity of the Universe by considering it in the guise of two abstractions—namely the World of multiple Activities and the World of coördinated Value. The prime characteristic of one World is change and of the other World immortality. But the understanding of the Universe requires that each World exhibits the impress of the other.

For this reason the World of Change develops Enduring Personal Identity.' (p. 693).

[2] PR, cf. p. 122. [3] PR, cf. p. 137.

Whitehead's account in P R of the relationship between God and the World is then capable of being translated into the language of the physical field. Instead of this abstract scheme having as terms 'events pervaded by perceptual characters', the terms would be 'events characterized by lines of force' (or electro-magnetic characters). Indeed his whole account of the creative advance into the future seems rather to be based on electro-magnetic theory than on the passage of nature observed in our everyday experience. The latter has not the same idealized smoothness or regularity about it. At times things around us seem to be at a standstill, at others we are aware of the world as full of rapid change. This phenomenon has not entirely passed unnoticed, and has led people to speak of psychological time.

Further, his account would seem to coincide almost exactly with his position in M C, where we had a certain polyadic, many-termed relation *R*, and the entities forming its field, the 'stuff' constituting the moving material world, were 'the lines of force of the modern physicist', which he took as 'the ultimate unanalysable entities which compose the material universe'. God, or the system of extensive relations, can then be compared with the polyadic relation *R*, the actual world, made up of a multiplicity of events pervaded by physical characteristics, with 'the ultimate unanalysable entities' or 'lines of force'. On his later view, however, the set of entities forming the field consists of events pervaded by physical (or sensory) characteristics. It is interesting to observe that Whitehead's notion of events as elements of becomingness (having the property of creative advance) bears upon it the vector character of 'lines of force'. There seems then ground for believing that he has modelled the former notion upon the latter, rather than modelling it upon the immediately given data in perception.

Consider in more detail Whitehead's account of 'creativity', that it 'is the pure notion of the activity conditioned by the objective immortality of the actual world'.[1] The exact interpretation of this notion seems to have puzzled some people. He seems, however, to be referring to the field of physical activity pervading space-time (or the structure of events). The field is conditioned by the objective immortality of the actual world, since as it streams into the future it is not only determined by the present event, but also by the events making up its past history. In this way there is a transmission of character. Or as he puts it, 'The whole spatial

[1] P R, p. 42.

universe is a field of force—or, in other words, a field of incessant activity'.[1]

We obtain some further evidence on this point in S M W, when he tells us that the laws which condition this field of electro-magnetic activity 'are nothing else than the conditions observed by the general activity of the flux of the world as it individualizes itself in the events'.[2] He goes on to give us a more detailed account of this 'general activity', asserting that 'Its attributes are its character of individualization into a multiplicity of modes, and the realm of eternal objects which are variously synthesized in these modes. Thus eternal possibility and modal differentiation into individual multiplicity are the attributes of the one substance.'[3]

V

Let us examine Whitehead's conception of God as the principle of concretion[4] that limits events to having a serial order, as a logical function limits values to having that defining form. As we have seen, Whitehead attempts to formulate the conceptual structure of experience in purely logical terms. He does this by replacing certain of its general features by logical entities, variables taking the place of the basic qualitative elements. When dealing with direct experience in terms of this scheme, we can think of the variables as having been replaced by particular values or sensory experiences.

This is undoubtedly a neat way of putting it, if there is such a thing as a logical scheme of order underlying our experiences. Otherwise, apart from its use as a methodological device, there seems little reason for this procedure, a view to which Whitehead leans in his discussion of philosophic method. Even if the order in experience is an attenuated one, for which such a network of logical relations has been substituted, it gives us no guarantee, as Whitehead sometimes seems to imply, that future events will have the same order as those in the present.

When he talks of the 'Primordial Nature of God', he would then seem to be referring to an idealized network of logical relations, which he tries to fit on to the universe of happenings. To suppose that this abstract system is actualized by events in the same way as a value satisfies a function, is no doubt a convenient shorthand way

[1] N L, p. 27.
[2] S M W, p. 190.
[3] S M W, p. 220.
[4] Cf. P R, p. 488 and S M W, Chapter XI.

of symbolizing the physical (or perceptual) process as it advances into the future. It is certainly not a fact about the world.

Further, Whitehead's conception of an object as a historical route of events, i.e. a set of serially related events, is certainly a most sophisticated notion and seems to be modelled on the geometric construction of a line; i.e. a set of points ordered by a serial relation. It would therefore appear to be much more a method of logical representation, than a description of the perceptually given. On the other hand, his notion of the creative advance seems in some ways to be modelled on the physical concept of the electro-magnetic field. However, we are not immediately aware of events, but of something which approximates more to the Gestalt notion of figure and ground. We should therefore not allow ourselves to be dictated to either by the foundations of geometry or by physical theory in our descriptions of direct experience.

Whitehead's whole account of the antithesis between God and the World seems to be based on a bifurcation.[1] He first states that the world has a structure. Now though experience has some systematic character, the structure Whitehead talks about does not seem to be obtained so much from direct experience, as by postulating hypothetical relations between portions of the perceptual field. Apparently he takes it for granted that everyone will accept his statement that we live in a universe of events, just as people take it as an everyday matter of fact that they live in a world of trees, houses, and other objects. This attempt to re-educate us to believe that objects are really tied-up groups of events fails, at least in this case, because as a theory it is not rooted in our direct experience, but rather in a certain logical technique, the notion of an ordered series. The latter notion may be derived from direct experience, by a process of substituting for certain of its features a system of precisely defined logical entities. But this does not mean that it exists in that form in nature any more than a mathematical point does.

It would also be interesting to know in what way God and the World stand over against each other, or to put it symbolically, *R*(. . .) and *A*, *B*, *C*, (i.e. the multiplicity of events). Is Whitehead talking of the relations and terms standing over against each other as symbols do on a piece of paper? Does it mean anything to speak of God or *R*(. . .) as seeking physical multiplicity?[2] It is like saying that a logical function strives to obtain a 'satisfaction'. Further, we

[1] Cf. P R, chapter on 'God and the World'. [2] P R, cf. p. 493.

fail to understand how 'the atomic world of a multiplicity of events' can seek a perfected unity;[1] i.e. seek relations to order it into the familiar world of enduring objects. Whitehead in his account of the interplay of opposites is, we think, largely manipulating abstractions. He is doing, perhaps unwittingly, the same sort of thing Lewis Caroll did in *Alice In Wonderland*, clothing abstract notions in concrete terms applicable to human experience.

VI

It is interesting to see how Whitehead links up his assumption that such an abstract relational system underlies the universe with our belief that there is an order in nature which gives us some justification for making forecasts as to future occurrences. We find him pointing out in S M W that the faith in the order of nature which has made possible the growth of science is a particular example of a deeper faith which cannot be justified by any inductive generalization. He tells us that it springs from direct inspection of the nature of things as disclosed in our own immediate present experience. 'To experience this faith', he tells us, 'is to know . . . that detached details merely in order to be themselves demand that they should find themselves in a system of things.'[2]

Whitehead is therefore of the opinion that our instinctive belief in an order in nature (though it apparently cannot be justified by any inductive generalization) is to be traced to a deeper faith which has its origin in our direct inspection of immediate experience. We are told what the experience of this faith consists in: namely, we know that the fragmentary perceived data must be related together with the rest of the universe into one ordered system. But, even suppose we admit Whitehead's contention, are we given this knowledge in direct experience? At the most it could only arise when we reflect upon the perceptual data.

Whitehead further would seem to identify (i) the faith that the universe has such a systematic character with (ii) knowledge that it actually has. One thing is quite plain, (i) does not entail (ii). Such a faith might no doubt spring from our inspection of the present experience, and as a result of seeing that it has a certain relational character we might go on to believe that the future will also be ordered by such relations. But we certainly do not know that this will be the case for future events, nor do we know that the

[1] P R, cf. p. 493. [2] S M W, pp. 23-4.

perceptual elements in order to be themselves demand that they should find themselves in such a system.

One can see how little Whitehead is really appealing to the direct inspection of things, when he supposes we know that 'the harmony of logic lies upon the universe as an iron necessity'.[1] Whitehead's reason for believing in the order of nature is not, however, based on purely empirical grounds as he would seem to indicate, but rather upon certain assumptions borrowed from mathematical logic and applied to the field of experience. Of course, we are not suggesting that we cannot arrive at such a logical scheme by a process of substituting for the crude data of experience more refined logical entities. Indeed in A E[2] Whitehead clearly recognizes this by stating that the uniformity which must be ascribed to experience does not belong to the immediate relations of the crude data of experience, but arises as a result of such a process of logical substitution. In other words, it has a second-order character.

As we have seen in Chapter I, Whitehead seems to push his earlier view into the background in S M W, when he tells us that this pattern of relationships, or deductive superstructure, is imposed alike on external reality and on our abstract representations of it. As he puts it, 'It means that for things to be together involves that they are reasonably together'.[3]

Now this involves a large assumption. What is the criterion by which we determine how things are reasonably together? Does this mean anything more than that the elements in experience must have certain logical characteristics, that each entity has a certain individuality and is also related together with other entities in one system? Only in this way can his view that the necessities of abstract logic are presupposed in interrelated experience make any sense.[4] Further, what grounds are there for supposing that the conceptual structures we contemplate in thought are also imposed on the physical world? Whitehead would probably hold that, unless our experience had some logical form, it would not be possible to handle it, and systematic knowledge would be out of the question. However, apart from the fact that we are probably aware of certain crude analogues in perception which we replace in thought by the corresponding precise logical notions (such as identity, non-contradiction, excluded middle,

[1] S M W, p. 24.
[2] A E, cf. p. 245.
[3] S M W, pp. 34-5.
[4] S M W, cf. pp. 33-4.

etc.), this view overlooks the Kantian solution that these logical notions may be largely intellectual instruments.

VII

We should perhaps enquire into Whitehead's analysis of the nature of religion, and see what role the order of nature plays there. Whitehead's contribution would seem to be rather to the psychology (or sociology) of religion than to Natural Theology. He is concerned with pointing out these general elements of order in the universe, which he believes give rise to our particular religious emotions and religious forms of behaviour. As he says in R M, his aim was to give a concise analysis of the various factors in human nature which go to form a religion, directing our attention to the foundation of religion on our apprehension of those 'permanent elements apart from which there would be no changing world'.[1]

Whitehead therefore (i) analyses the factors in human nature which go to make up a religion: the particular emotions, thoughts and forms of behaviour; (ii) indicates that it is founded on our apprehension of certain permanent features in the world; and (iii) argues that the particular form taken by a religion is dependent upon our psychological reaction to these elements of order in nature.

He brings this out further by telling us that, though the concept of God is one essential element in religious feeling, the converse is not true; 'the concept of religious feeling is not an essential element in the concept of God's function in the universe'.[2] In other words, he thinks our apprehension of this system of order (i.e. of the permanence amid the flux of change) to be an essential element in religious feeling. By this he no doubt means that our particular emotions and thoughts are, as it were, interwoven with the experience of eternality which arises when we contemplate these elements of order in nature. But, on the other hand, though religious feeling is concerned with our emotional reaction to this order, it is not essential to our apprehension of it. Whitehead therefore recognizes the subjectivity of the particular religious emotions, and the social character of the associated forms of behaviour. He can therefore say that the principles of religion may be eternal, but the expression of these principles is usually in terms of the imaginative picture of the world entertained in

[1] R M, Preface. [2] P R, p. 294.

particular ages and among particular peoples. Whitehead believes that religion is the translation of general ideas into particular thoughts, particular emotions, and particular purposes. As he tells us, 'Religion is an ultimate craving to infuse into the insistent particularity of emotion that non-temporal generality which primarily belongs to conceptual thought alone'.[1]

What Whitehead seems to be doing is to point out that these general logical features which we discover in nature are taken by religion and interwoven with particular human emotions such as love, fear and awe. The precise form this feeling takes, together with its associated thought and patterns of behaviour, depends upon the particular epoch and society in which the individual finds himself. It is clear that the kinds of experience associated with different religions vary enormously. But Whitehead would claim that they are all centred round the same general notion of order (or permanence) with its suggestion of some kind of necessity or Law in Nature. In this way these elements of order are translated into particular emotions, thoughts and purposes, which may differ from each other as much as the primitive African religions differ from those of our sophisticated western civilization. They may be anthropomorphized in the form of the Greek Parthenon, or take on an animistic turn, as, for example, in the Egyptian deities.

Whitehead is of the opinion that from the belief that the universe was subject to Divine Law sprung the unquestionable faith in the reign of Natural Law. The Greeks, for example, portrayed in tragedy the inevitableness of destiny as it worked itself out in the actions of the heroic characters. 'Fate in Greek Tragedy becomes the order of nature in modern thought.'[2] 'The laws of physics are the decrees of fate.'[3] This attitude of rationality (and system) also manifested itself in the abstract principles of Roman Law, which in its turn influenced the mentality of the Middle Ages.

Whitehead further argues that the belief that every detailed occurrence can be correlated with its antecedents in a perfectly definite manner comes from the medieval insistence on the rationality of God, who personally supervised and ordered every detail in the universe. His conclusion then is that 'the faith in the possibility of science, generated antecedently to the development of modern scientific theory, is an unconscious derivative from medieval theology'.[4] Science then owes something to theology, namely the general presupposition that the world is subject to law,

[1] P R, p. 21. [2] S M W, p. 12. [3] S M W, p. 13. [4] S M W, p. 16.

and this, he would argue, forms the conceptual framework in which scientists have instinctively worked. Though scientists may not have worried very much about it, their investigations have shown that the belief in the order of nature was at least justified pragmatically. How Whitehead's view of God differs from that held in the past by religious thinkers can be seen from his answer to the following question. He asks, 'What is the status of the enduring stability of the order of nature?'[1] The tendency, he points out, has been to evaluate nature summarily in terms of some greater reality standing behind it. 'This reality occurs in the history of thought under many names, The Absolute, Brahma, The Order of Heaven, God.'[2] According to Whitehead, then, these permanences found in nature have in the history of thought usually been connected with some principle standing behind our actual world. But he does not want to jump from such an order to the belief that there is some supernatural being imposing it. His aim is to discover whether nature does not in its very being contain its own conditions, whether this order does not arise from the relations between things themselves. Otherwise, he thinks the appeal to a supernatural being tyrant-like imposing an order upon nature is the refusal of rationality to assert its rights, namely to analyse experience, and to see whether elements can be found within it explanatory of this fact.

SUMMARY

It is argued that the general system of extensive relationships, or extensive continuum, is identical with the 'Primordial Nature of God', having all its logical properties. Whitehead thinks of it as the ordering entity in nature since it relates the multiplicity of events into a well-ordered system. Sample passages illustrating the way Whitehead uses the notion of God in his philosophy are examined. It is seen from these that the notion of God has two aspects: (i) a logical aspect—as an ordering relation *R* unlimited in its range of possibilities, and (ii) a physical aspect as relating the multiplicity of events into our common world. Whitehead also indicates the factors in human nature which go to make up a religion—the particular emotions, thoughts and forms of behaviour which arise from our apprehension of these permanent features of order in the world.

[1] S M W, p. 115. [2] S M W, p. 115.

CHAPTER IV

THE REALM OF ETERNAL OBJECTS

I

MOST commentators assume that there are two main stages in Whitehead's philosophy; the first is supposed to cover the period of his nature philosophy, the second the metaphysical stage with its leaning towards Platonism. In our opinion such a disjunction does not really exist; the principles underlying his later works are rather a development of those to be found in his earlier writings.

Take perhaps one of the most interesting and at the same time puzzling concepts of his later stage, 'the realm of eternal objects', a doctrine which makes his philosophy particularly tempting to a Platonist, numerous analogies having been drawn between it and Plato's realm of forms. Nevertheless, it is difficult to understand how this interpretation can be reconciled with the professed empiricism of Whitehead.[1]

The primary aim of this chapter will be to show that Whitehead was not putting forward a Platonic realism, but was rather applying certain concepts from symbolic logic, for example, the concept of the propositional function and the variable, to elucidate the structure of experience. In this connection, we should perhaps say straight away that by the realm of eternal objects Whitehead is really referring to an abstract logical structure derivative from the relation of extensive connection holding between events.

According to Whitehead, the world given in perception is made up of events which extend over each other to form by their overlapping a spatio-temporal system. Every event, no matter when and where it appears, must be thus related to other events. Otherwise there would be no continuity in nature, no past, no future, only a multiplicity of disconnected events. Hence for Whitehead the relation of extension holds for all events and epochs. He

[1] P R, Preface, v., cf. his assertion that: 'The writer who most fully anticipated the main positions of the philosophy of organism is John Locke'.

symbolizes it in the form aKb, where a may be the event 'Jones eating his breakfast' and K its extension over the event b, 'Jones eating his porridge'. In this sense, extension taken by itself is not something actual, it is not an event nor made up of events, but is a relation between them. In abstraction from events it is a mere potentiality, a logical framework in which events can be inserted just as values can in a function.

Hence, when Whitehead refers to the realm of eternal objects, he is not dealing with particular qualitative characteristics (or even class concepts, such as redness or greenness), but with groups of unspecified entities in definite patterns of connection. He is interested in the logical structure of our experience rather than its qualitative detail. According to Whitehead, these extensive relationships are expressible without reference to any particular colour, sound, etc., just as a propositional function ϕx may be expressed without reference to any proposition.

And this is what Whitehead has in mind when he tells us 'Accordingly the relationships (as in possibility) do not involve the individual essences of the eternal objects; they involve *any* eternal objects as *relata*, subject to the proviso that these *relata* have the requisite relational essences'.[1] That is to say, the extensive patterns in the abstract framework do not involve specific sensory characteristics, but refer to any such characters provided they have the requisite structure. In contrast to this, when these abstract forms are given a determination (or value), they are transformed into events, into definite spatio-temporal patterns of colours, sounds, and other sense-qualities.[2]

When Whitehead, therefore, speaks of the realm of eternal objects, he is not indulging in an Aristotelian classificatory analysis in terms of species and genera. He believes that the classificatory method shatters the universe into a bundle of fragmentary qualities and leaves out the essential relatedness of things. Whitehead introduces a new technique in the analysis of fact. Instead of using the classificatory method (which is only a half-way house, for it merely deals with class-concepts of the type $\phi(a)$, e.g. any object belonging to the sort red), he replaces it by the complete

[1] S M W, p. 205.

[2] If x, y, z stand for any set of sense-objects, and R for the extensive relation, then when we determine $R(xyz)$ by assigning values to the variables we get a specific sensory pattern; x, y, z are now sensory characters and R is a definite spatio-temporal pattern.

analysis of the form 'any object in such a pattern', i.e. $\phi\hat{x}$ or $\hat{x}R\hat{y}$. In this analysis of an event, we replace 'the specific spatio-temporal pattern of colours, sounds, etc.' by 'a set of variables in a definite logical pattern'. Hence, Whitehead in his realm of eternal objects is primarily concerned with extensive forms in abstraction from any particular sensory structure; and just as a sensory structure can be analysed into a spatio-temporal pattern and specific sense-objects, so this abstract structure can be said to consist of a logical form, and variables representing any objects. Whitehead is engaged upon an examination of the logical structure of experience.

This links up directly with his analysis of an eternal object as having (i) an individual essence, and (ii) a relational essence. In (i) we deal with the eternal object's specific characteristic, such as colour (or shape),[1] which may exhibit itself in some perceived sensory pattern. On the other hand, (ii) is its pattern of connection, which exhibits itself in the event as a definite spatio-temporal pattern (an individual shape enduring through a lapse of time) but which, considered abstractly, merely has the character of a logical form, holding for a variety of sensa.

We can now see what Whitehead means when he informs us that eternal objects are 'isolated' in the realm of possibility 'because their relationships as possibilities are expressible without reference to their respective individual essences'.[2] Whitehead considers the eternal objects defined by this scheme to be 'isolated', since as abstract possibilities they are divorced from actual fact. But nevertheless we are enabled by means of their logical relationships to express them as abstract possibilities without needing to refer to any specific sensa, just as we can in the case of a propositional function express a range of possible propositions without needing to refer to any specific one. On the other hand, when Whitehead says that in an event there is 'a real togetherness of their individual essences', he means that these specific qualitative characteristics are related together in the form of a definite spatio-temporal sensory pattern.[3]

[1] Whitehead also applies the term 'eternal object' to qualities such as shape. It should be noted that the 'individual essence' is also abstract, since we deal with such class concepts as red or green. Only their particular instances are to be found in perceptual fact.

[2] S M W, p. 205.

[3] Or as he puts it in S M W, p. 130, it is 'a grasping of diverse entities into a value by reason of their real togetherness in that pattern'.

Incidentally, Whitehead is covering the same sort of ground as he did in M C, where he attempted to formulate the conceptual structure of physics. 'Consequently, the paper may be said to broach the idea that physics is "one application of a logical system . . . the existing science being called upon to supply the entities and relationships among entities which substituted for the variables, give the abstract structure its important interpretation".'[1] What Whitehead is trying to do in S M W is to exhibit the logical component of experience in the form of such an abstract structure. In this case experience may be said to supply the values, i.e. the events (the specific sensory characters and the specific spatio-temporal relationships between them) which when substituted for the logical variables, give this abstract structure its determination, transforming it into a systematic complex of perceptual objects related in space and time.

II

The most complete account of the realm of eternal objects is to be found in the chapter 'Abstraction' in S M W. We are told there that each eternal object such as *A* has a 'relational essence',[2] i.e. a set of extensive relations which give it a status in this abstract system. It should be noted that the symbol *A* stands for any specific type of eternal object (or individual essence), it might be blue, an acid taste, or a nauseating smell.[3]

Whitehead goes on to state that the relationships *A* has to other objects in this realm, e.g. *B*, *C*, *D*, etc., are internal, since they stand determinately in its relational essence. By this he means that these relationships form its abstract pattern of connection which gives the object *A* its standpoint in this system, and from which it cannot be abstracted without losing its significance. As a

[1] Schilpp, *Philosophy of A. N. Whitehead*, p. 45. (Lowe, 'Development of Whitehead's Philosophy'.)

[2] S M W, cf. p. 198.

[3] We may be asked to say plainly what Whitehead means by an eternal object. Is it a quality, a relation or what? In S M W, for example, an eternal object is a double-barrelled concept. When he talks about its 'individual essence' he refers to such abstract class concepts as colours and sounds, which taken in themselves are devoid of perceptual content. In P R they are called eternal objects of the subjective species. On the other hand by its 'relational essence' he refers to logico-mathematical relationships. In P R they are called eternal objects of the objective species.

variable *A* is indeterminate and presupposes a structural context, and out of its context it is meaningless. On this abstract plane we cannot speak about the entity *A* in itself, but only *A* as in its relational context, e.g. *R*(*ABC*), and in its turn this structure cannot be divorced from the rest of the system to which it is related.

This brings us to consider in more detail the precise connection between the spatio-temporal continuum and the realm of eternal objects. He states that 'the spatio-temporal relationship, in terms of which the actual course of events is to be expressed, is nothing else than a selective limitation within the general systematic relationships among eternal objects'.[1] But, it may be asked, how can the spatio-temporal continuum made up of particular events be a selective limitation within what seems to be merely an abstract framework? The answer is simple—it is a selective limitation in the sense that it can be thought of as a particular value given to this abstract scheme. Instead of dealing with a system of extensive relations, we now deal with specific types of order spreading themselves spatially and temporally.

Whitehead explains that by ' "limitation", as applied to the spatio-temporal continuum, I mean those matter-of-fact determinations—such as the three dimensions of space, and the four dimensions of the spatio-temporal continuum—which are inherent in the actual course of events, but which present themselves as arbitrary in respect to a more abstract possibility'.[2] They are the content or matter-of-fact determinations given to the abstract scheme by the actual course of events. But these determinations are arbitrary in respect of other possibilities (i.e. values) defined by this scheme. The spatio-temporal continuum is merely one of the types of order, or possible values, this conceptual scheme could be given.

It will be seen that by 'the general systematic relationships among eternal objects' Whitehead is really referring to the extensive relations which, he assumes, underlie every epoch—the fact that events, no matter when and where they appear, must be thus ordered or extensively connected. Hence, this unlimited extensive scheme can be said to be limited by events which reduce it to the four-dimensional continuum. The difference between these two schemes of order will readily be noted. Whereas the extensive scheme, owing to its complete generality, defines an

[1] S M W, p. 200. [2] S M W, p. 200.

infinite range of possible types of order, the spatio-temporal continuum may be described as a selective limitation, since it has certain definite characteristics, e.g. three dimensions of space and one of time.

Hence, when Whitehead asserts that every event is a limitation imposed upon possibility, he means that the events determine, or give a specific character to, these abstract sets of relationships so that they appear in space and time in the form of perceptual objects. This explains what he has in mind when he tells us that by virtue of this limitation a particular value arises; and by 'value' he means a specific sensory pattern, and not (as might be thought) a peculiar feeling of aesthetic appreciation in the mind of some observer.

When in perception we are aware of a sensory quality, it has to appear with other sensory characters in a spatio-temporal context, i.e. at a definite place and time—to take an example, the specific pattern $R(ABC)$ which may be three colours related together in the form of a triangle. As a possibility, we could imagine it as having any kind of exemplification—it might be a triangle observed by a Pythagorean in the sixth century B.C., or one drawn in crayons by a modern child. But when we are aware of this pattern now, it is restricted to appearing in just this place and time: we may, for example, see a coloured triangle in front of us.

A word should be said about Whitehead's doctrine of 'being' and 'not-being', which with memories of the *Sophist* has been hailed as an example of his Platonism. It is, however, only a further elaboration of the above account of the inclusion and exclusion of specific characteristics from our immediate field of perception. We need to note that an eternal object A is 'not-being' for him when it is regarded merely as an abstract possibility. In this sense an object A, say red, is 'not-being' with respect to a definite event, when it is excluded from our immediate field in any of the possible sensory patterns it could enter in as an element.

On the other hand, red is considered as 'being' in respect to a definite event when we are aware of it in one of the possible sensory patterns it can enter into as an element. Consider, for example, the red book now on our table. We may say that red is here included as 'being', but when we look out of the window and see the green lawn, then for that event it can be considered as 'not-being', as at that particular moment we may not be aware of any object in which red appears.

III

As we have seen, when Whitehead discusses the nature of the realm of eternal objects, he states that every part of this scheme stands in internal relations to every other part. He now draws our attention to a difficulty which arises in respect to internal relations. Apparently, on such a view, everything must depend upon everything else. 'But if this be the case, we cannot know about anything till we equally know everything else.'[1] Whitehead, however, thinks this supposed necessity to be quite untrue. It is clear that, with perhaps the exception of some Idealist philosophers, most of us do recognize that we can assert particular truths about particular things. He therefore has 'to explain how there can be internal relations, seeing that we admit finite truths'.[2]

Now this difficulty also faced him in P of R, where, after putting forward his argument that there is a systematic relatedness in nature, he formulated the same question in rather different terms. We are told that, if a factor (i.e. an event) is not itself apart from other factors, then to express any truths about one entity you must take into account its relations to all other entities. However, if this is the case, the attainment of finite truths is beyond us.[3] Though the problems discussed in both S M W and P of R appear to be identical, they nevertheless differ in their degree of generality. In one case the internal relations belong to the abstract extensive scheme—in the other to the uniform spatio-temporal continuum.

To understand Whitehead's point, let us look at the question more concretely. The existence of such limited relationships in the spatio-temporal continuum arises from its divisible character, divisible because it can be analysed into the particular spatial boundaries and definite periods of time of our common-sense world. We cannot help being struck that there are such limited structures in our perceptual field standing over, as it were, against a common background. As Whitehead puts it, 'We endeavour to lift into consciousness meaningful units, such as the whole picture, the whole building, the living animal, the stone, the mountain, the tree'.[4] In this way, we are enabled to attend to particular sensory groupings and to state particular truths about them, without having to take into account the rest of the universe.

In Whitehead's account in S M W these structures are lifted from

[1] S M W, p. 203.
[2] S M W, p. 203.
[3] P of R, cf. p. 22.
[4] M T, pp. 169-70.

their perceptual context and are, as it were, held up in their naked abstractedness for our inspection. We are told there that, 'Since actual occasions are selections from the realm of possibilities, the ultimate explanation of how actual occasions have the general character which they do have, must lie in an analysis of the general character of the realm of possibility'.[1] In other words—since on Whitehead's view events can be considered as selections from the range of possible events defined by this abstract scheme, any ultimate explanation of how they have finite relationships and are yet mutually related to each other in nature, must lie in an analysis of the extensive scheme. The reason for this apparently perverse procedure arises from the fact that, since the extensive scheme is the conceptual counterpart of the spatio-temporal system given in perception, every event must conform to its defining conditions. It follows, therefore, that statements about the general character of the extensive scheme are at one and the same time statements about the general conditions satisfied by events.

The primary metaphysical truth concerning this realm is, Whitehead thinks, its analytical character. By this he means that we can analyse the relationships any eternal object *A* has to other objects in this scheme into an indefinite number of limited structures. For example, if *B* and *C* are two other eternal objects, there is some perfectly definite relationship *R*(*ABC*) involving only these three objects, and there may also be more inclusive patterns including this structure as a component, e.g. *R*[(*ABC*)*DF* . . .], etc. This may best be illustrated by expressing it in perceptual terms, e.g. *R*(*ABC*) may be three definite colours related in the form of a coloured triangle, or it may be an element in a wider pattern, say the sign outside a YMCA hostel.

When Whitehead therefore asserts that the realm of possibility provides a uniform scheme of relationships among finite sets of eternal objects, he would seem to mean that the abstract scheme provides such a system of relationships among finite extensive patterns of the type *R*(*ABC*), *R*[(*ABC*)*DF* . . .], etc. The full significance of his statement is best seen when we look at it from the point of view of perception. Here the spatio-temporal scheme may be said to connect or separate the sets of sense-data (or perceptual objects), each perceptual object having a definite location or boundary. The abstract extensive patterns are then to be thought of as the logical forms of these perceptual objects.

[1] SMW, p. 203.

He formulates the principle for the existence of such limited relationships in an internally related system in the following manner: 'The whole principle is that a particular determination can be made of the *how* of some definite relationship of a definite eternal object A to a definite finite number n of other objects, *without* any determination of the other n objects, $X_1, X_2, \ldots X_n$.'[1] That is to say, we can indicate the way in which A is related to a finite number of other objects in this scheme without specifying what exactly these objects are. In this way, we can isolate certain finite structures without needing to refer to the rest of the related scheme.

The above, though stated in more general terms, is almost identical with the answer given by Whitehead to a similar question in P of R. He told us there that in order to perceive A (some definite entity) we do not have to be conscious of the related entities B, C, D, with cognisance by adjective,[2] i.e., it is not necessary to be aware of their specific colours, sounds, etc. We only need to know that it has these uniform relations to them,[3] in other words, that they are related together in one system. For example, we may be aware of the brown of the table in front of us although remaining ignorant of the colour of our next-door neighbour's lawn-mower, to which it is spatio-temporally related. Whitehead's reason for postulating a uniform spatio-temporal continuum is that, if we suppose there is such a uniform scheme of relationship which includes and stretches beyond our limited perceptual field, we can still know that there are events beyond it though we may not know what specific characteristics they display. We can, for example, specify a time and place on the other side of the moon, despite our ignorance of what is going on there.

Faced with the need for justifying induction, Whitehead believes that, unless events were uniformly related to each other, induction and for that matter systematic knowledge would not be possible. To take a trivial example; as we sit before the fire we may hear a 'miouw' and assume that there is a cat in the next room. But as Whitehead points out, we would not be able to make this inference unless the 'miouw' and the cat are related in the same manner in the next room as they usually are in this one, unless we could thus

[1] S M W, p. 204.

[2] P of R, cf. p. 23. Cognisance by adjective bears some resemblance to the notion of 'individual essence'.

[3] I.e., cognisance by relatedness. This resembles the notion of 'relational essence'.

assume, as we do in everyday life, that the behaviour of objects remains unaffected by a mere change in spatio-temporal position. Therefore, Whitehead argues, unless this ground of uniformity existed, knowledge of events outside our own perceptual field and hence induction would be impossible.

To sum up, Whitehead in S M W gets over the difficulty inherent in the concept of finite internal relations in the extensive scheme by two principles, (i) that *A*'s relationships to other objects in this scheme merely involve these objects as bare relata, or variables, without reference to their sensory characteristics; (ii) that since *A*'s finite perspective places it as one entity amid a definite system of entities, this general scheme of relationships is divisible into a multiplicity of finite perspectives. It will be seen that (ii) depends on (i), for to be acquainted with *A* is to know that it is related as an element in such a system; and according to (i) we can have this knowledge though remaining ignorant of the characters of the other related entities.

To take a concrete illustration: if when climbing a mountain we were suddenly confronted by a bear, all that would concern us at that moment would be the fact that he is located in front of us now (i.e. that he has a definite finite standpoint within the spatio-temporal system). In order to be fully conscious of the bear, we certainly do not need to know that he is spatio-temporally related to the planet Mars, or for that matter to anything else in the universe. But what Whitehead would say is that, although nature is analysable into a multiplicity of finite structures, bears, trees, men, etc., which have their own individuality, these objects are nevertheless related together in the one system.

IV

Whitehead states that the analysis of the general scheme of relatedness exhibits it as a multiplicity of complex eternal objects. As there will probably be some difficulty in grasping how the realm of eternal objects is thus analysed, our best plan would be once again to regard it in its restricted perceptual form. When we examine our perceptual field, we see that it is made up of well-defined entities, such as houses, trees and stones, which are analysable into sets of sensa. If we now consider the general scheme without reference to events, it may be said to consist of a multiplicity of extensive patterns which are, as it were, the abstract form of their more concrete counterparts.

But, it may be asked, how does this fit in with Whitehead's statement that a complex object is a 'finite relationship made up of a limited set of component objects'? This difficulty is easily met. Complex objects play a dual role for Whitehead. In the spatio-temporal scheme they are perceptual objects analysable into limited sets of sensory characters; as elements in the abstract scheme they are extensive structures, the components of which have no sensory content since they are variables. And, as in the case of perceptual objects, these abstract patterns may be of varying grades of complexity. Hence, when Whitehead refers to the analytical character of this system, he means that it is analysable into a variety of such extensive structures. Unless the reader grasps that these objects have both a perceptual and a logical form, Whitehead's position becomes difficult to understand.

The fact that the realm of eternal objects is a system of related structures, facilitates its analysis into grades of complexity. In the lowest grade (that of zero complexity) he places objects whose specific characters are simple (e.g. a definite shade of green). He considers next any set of such simple objects having a definite relationship to each other, say $R(ABC)$, which to give his own illustration may be 'three definite colours with the spatio-temporal relatedness to each of three faces of a regular tetrahedron, anywhere at any time'. Hence, $R(ABC)$ (as a set of elements combined by a triadic relation) is an eternal object of the lowest complex grade. And, as will readily be seen, analogously there are structures of successively higher grades of complexity (or degree), just as in our perceptual field some objects are more complex than others.

If we now take any complex object, e.g. $S(D_1, D_2, \ldots D_n)$, then the individual characters of the component objects $D_1, \ldots D_n$ are to be considered as constitutive of the total character of this whole pattern. Putting it in perceptual terms, the complex character of some definite perceptual object may be said to be a composition of the specific sensa into which it is analysable. And, as the whole object is more comprehensive than any of its component elements, it is hence a grade above their highest grade of complexity. The realm of eternal objects may therefore be analysed into simple objects and into various grades of complex objects, much in the same way as we can analyse our perceptual field into houses, trees, etc., and stray sensa such as sounds and smells, which do not form part of any definite object.

Whitehead terms any such complex object having a set of simpler elements subsumed under it an 'abstractive hierarchy'. A hierarchy is termed 'finite' if it includes a limited number of such relationships; this is the case with the structures appearing in finite sections of our perceptual field, and in mental phenomena such as dreams, images, etc. It is considered 'infinite' if it includes objects of all degrees of complexity, for example, when we deal with a spatio-temporal perspective which stretches indefinitely beyond our field of perception. In this case 'we are aware of the distant environment fading away in the general knowledge that there are things beyond'.

Whitehead then conceives an 'abstractive hierarchy' as a complex structure containing within itself sets of structures of successive grades of complexity (simple and complex eternal objects). We may better bring out the meaning of this notion by translating it into perceptual terms. When we observe, say, a pleasant country landscape, it is analysable into a complex of related entities, trees, hills, fields, etc. (or rather, overlapping events in which such objects are implicated), which can be further analysed into a collection of sensory characters, colours, sounds, smells, etc. These sensa may be said to be the basic elements related within this structure.

This property of an abstractive hierarchy, that it defines a set of subordinate structures, is called by Whitehead the 'condition of connexity'. By this he means that at any grade of complexity of such a hierarchy we can find a pattern which includes (or extends over) a set of structures of a lower grade than itself. Hence it follows that, with the exception of the grade of simple objects, every structure within this abstract perspective includes other structures and is itself included by other more comprehensive patterns.

We may illustrate this again by taking a perceptual example. When travelling in a train from, say, London to Edinburgh, we may look out of the window and observe standing over against the horizon a clump of trees perched on a hill top. In this case we could say that a certain set of sensa possessing peculiarities of colour and shape are related as components of the sensory pattern which forms the clump of trees. The clump of trees is itself included within the more comprehensive pattern which is the wooded hill, which in its turn stands in similar relations to the rest of the surrounding countryside.

Whitehead amplifies his statement by noting that the components of a complex eternal object are necessarily of a lower grade than itself, i.e. they have a lower degree of relatedness than the complex pattern in which they are included as elements. Hence, any object in such a hierarchy of the first grade of complexity (or degree of relationship) can have as relata only simple eternal objects, those of the second grade only structures of the first grade and simple objects, and so on for higher grades. In other words, a pattern having a high degree of connectedness, say R_n, can only include as elements complex objects of a lower degree than itself, i.e. R_{n-1}, R_{n-2}, R_{n-3} and the grade of simple objects. Hence R_n can be said to include (or extend over) this set of subsidiary structures. An abstractive hierarchy may then be thought of as defining such a series of structures, each structure including other structures and being included by other structures. The relationship holding between the successive grades is then a transitive one, i.e. if *A* is included in *B*, and *B* in *C*, then *A* is also included in *C*.

It will be seen that this 'condition of connexity' which an abstractive hierarchy has to satisfy (i.e. that every pattern of a higher degree includes patterns of a lower degree than itself) resembles the fundamental relation of extension, or whole and part, which Whitehead assumes holds between events. As he points out, a duration, i.e. 'the whole of nature apprehended in our immediate observation', can be discriminated into an inter-related complex of events having the whole and part relationship. In other words every event extends over events and is extended over by other events.[1] An infinite abstractive hierarchy would then seem to be the abstract structure of such a perceived duration.

V

This brings us to consider in more detail what Whitehead has in mind when he talks of a finite abstractive hierarchy. He tells us 'A finite abstractive hierarchy will, by definition, possess a grade of maximum complexity',[2] i.e. it is a complex structure whose degree of relatedness is greater than that of any of the structures it includes. Further, there must be only one such comprehensive pattern, as otherwise any structure having a lower degree of relatedness would not be subsumed under it. We need to note

[1] C N, cf. p. 59. [2] S M W, p. 209.

that the relationship of inclusion is asymmetrical, i.e. if R_n includes R_{n-1} then R_{n-1} does not include R_n.

Whitehead outlines his method for the analysis of finite abstractive hierarchies into simpler elements, as follows: any complex eternal object, 'defines a finite abstractive hierarchy to be discovered by a process of analysis'.[1] That is to say, if we start from what is really a finite extensive perspective, it is analysable into a set of structures of different degrees of complexity, just as a perceived perspective is analysable into a set of over-lapping sensory patterns.

Whitehead's account in its purely abstract form is rather difficult to follow, and because of this has probably been omitted by most readers of S M W. It can, however, be easily explained if we illustrate it by a specific example.

Let us imagine that we are standing on the battlements of Edinburgh Castle and looking down on the city; its streets, houses, open spaces, public buildings and factories stretch below us. The perspective which we now observe (or the vertex of the abstractive hierarchy R_n) is therefore analysable into a set of component structures, some of which are more complex than others. We begin our Whiteheadian analysis by selecting the most complex objects in the perspective, say the various readily distinguishable 'avenues' lined with buildings standing out in bold relief (i.e. objects of R_{n-1} complexity). For the moment, we will only consider these easily recognizable units and neglect simpler units such as isolated houses, etc.

At the second stage of our analysis, we analyse the 'avenues' into simpler elements, e.g. such clearly demarcated units as large blocks of buildings. These are, then, the most complex objects to be found at this stage (objects of R_{n-2} complexity). In our first analysis, however, we disregarded the simpler objects in the perspective, i.e. single cottages, blocks of flats and odd gasometers on the outskirts of the city, which is what we usually do in perception when we concentrate on the most important features of the field and assign to the background the structurally unimportant detail. We now return to them, and select those structures which are similar in degree of complexity to the most complex elements dealt with by our second analysis (i.e. $n-2$ relatedness), e.g. isolated blocks of flats and perhaps one or two factories.

We then, as it were, pool these similar structures (of $n-2$ related-

[1] S M W, p. 210.

ness) and proceed to an analysis of them, thus obtaining their components—simpler patterns such as individual houses which can be considered as the most complex patterns revealed by the third stage of our analysis (objects of R_{n-3} complexity). The previous procedure is then repeated, and the structures left over from the initial perspective of a similar grade (i.e. $n-3$ relatedness), e.g. single houses, clumps of trees, etc., are added to them. Continuing on these lines, we analyse the objects still further into simpler patterns, and finally, by such a series of successive analyses, reach the grade of simple sense-objects. The perspective has thus been reduced to its simple component elements.

We have thus a pyramid of grades of structure of decreasing complexity. The perspective from which our analysis started Whitehead terms the 'vertex', whilst the multiplicity, or group, of simple objects obtained by this analysis forms the 'base'.

The same method of analysis would, of course, be applicable to simpler patterns such as individual perceptual objects. A chair could e.g. be analysed into its components of a seat, back and legs, which could be further analysed into simpler patterns until we reach sense-objects. The number of stages in the analysis will vary with the complexity of the object. An elephant, for example, is more complex than a triangle, consisting as it does of such sub-patterns as legs, tusks, trunk, etc., which have a definite permanence of structure. The elephant taken as it were, *en masse*, is the connection of these simpler sub-patterns into a more comprehensive recognizable structure. Hence the complexity of an object (or its degree of relatedness) is to be graded in accordance with the sub-patterns it includes.

This seems to link up with the notion of societies in Whitehead's later works, where we are informed that nature is a complex of enduring objects, functioning as subordinate elements in a larger spatial-physical society. 'Also each of these enduring objects, such as tables, animal bodies, and stars, is itself a subordinate universe including subordinate enduring objects'.[1] Nature, for Whitehead, is then such a co-ordinated system of subordinate societies of different grades of complexity or levels. It is fairly evident that the notion of different levels of societies resembles that of different grades of complex objects. The main difference, of course, is that in the case of 'abstractive hierarchies' he is dealing with abstract extensive patterns, whilst in the case of societies he is

[1] A I, p. 265.

dealing with concrete patterns of events having a causal dependence upon each other.

VI

We need to consider Whitehead's view that any event has associated with it an 'infinite abstractive hierarchy' made up of a group of simple eternal objects. By this he means that every event can be thought of as exhibiting a specific perspective. In perception this takes the form of a spatial perspective stretching indefinitely beyond us, its various regions being illustrated by diverse sensory qualities. Whitehead calls this the event's associated hierarchy, and it is, he tells us, 'what is meant by the statement that it is impossible to complete the description of an actual occasion by means of concepts'.[1] As we have already pointed out, for Whitehead our knowledge of nature has an inexhaustible character about it. Thus, when we describe events, we know there are always things which lie beyond our immediate perceptual field and which are therefore not included by our immediate description. For Whitehead, each event can be said to be related to the rest of the spatio-temporal scheme, mirroring the universe, as it were, from its point of view.

Hence, in speaking of an infinite abstractive hierarchy included in each event, Whitehead is referring to its particular perspective of the spatio-temporal world. The group of objects which make it up consists of the sounds, smells, tactual data, etc., displayed in its various regions. Thus, for example, when we stand on Westminster Bridge and observe in the distance the bridges which span the river, what we really perceive is a perspective of these objects from our standpoint here in space-time.

The doctrine that every event has associated with it an infinite abstractive hierarchy resembles to some extent his earlier view that the percipient event has associated with it a present duration which has an infinite character, i.e. it is an unlimited whole. He tells us that the duration is primarily marked out by an interconnected display of sensa and of other associated objects.

Whitehead points out that our total experience also includes that which in cognitive experience takes the form of memory, anticipation, imagination and thought, which are also modes of inclusion of complex eternal objects. However, he adds, there is a

[1] S M W, p. 211.

difference between such mental states and our perceptions, since 'what is remembered, or anticipated, or imagined, or thought, is exhausted by a finite complex concept'.[1] In other words, when we remember, think or imagine, we are usually only aware of some particular memory image. As, for example, when remembering a visit we paid to the Sheldonian Theatre in Oxford, we have a mental picture of the building. Whitehead therefore asserts, 'the things apprehended as mental are always subject to the condition that we come to a stop when we attempt to explore ever higher grades of complexity in their realized relationships'.[2] In such cases we always find that we have thought of just this—whatever it may be—and of no more. In perception, on the other hand, our attention moves with ease from one object to the other. Hence we are not limited to the object which we may at that moment be specifically attending to.

The difference clearly shows itself if we contrast the material given in such mental states with that given in perception. In thought and imagination, for example, we have some selective control over the objects contemplated. When we wish to think of, say, the Sheldonian Theatre, we do so, though we may be a hundred miles away. We are thus enabled to contemplate some specific object apart from its spatio-temporal context. In perception, however, there does not seem to be this selective control over objects; we just accept them—they have a take it or leave it character. Further, though we are able to concentrate attention on specific aspects of the immediately observed field, there is always the spatio-temporal background to which it is connected.

According to Whitehead, then, 'an actual occasion is a prehension of one infinite hierarchy (its associated hierarchy) together with various finite hierarchies'.[3] Or, to put it differently, our total experience consists of one such spatio-temporal perspective whose regions are discriminated by diverse sensory qualities together with various mental experiences, images, etc. He now introduces a principle which he thinks is essential for the rational coherence of these different aspects of our experience, and calls it 'The Translucency of Realization'.[4] By this he means that any eternal object, in whatever type of experience it appears, whether in perception or imagination, will always retain its self-identity.

Whitehead contends that it is because of this principle that we

[1] S M W, p. 213.
[2] S M W, p. 213.
[3] S M W, p. 213.
[4] S M W, cf. p. 214.

are able to be aware of the same object as exemplified both in perception and in our subjective experience. For example, we can imagine red, and also at the same time check our image against our direct perception of the colour. It is this he thinks which makes the correspondence theory of truth possible. When we, as it were, place in juxtaposition, say, our memory image of red and our perception of it, the one-one correspondence that then exists between them is due to the same eternal object being included in both cases.

From the internal evidence of the text it would seem that Whitehead probably thought in terms of images, as he often uses the terms 'thought' and 'cognition' where we would use 'imagining'. If this is so, and it is not merely due to clumsiness in exposition, then Whitehead, when he argues that there is such a one-one correspondence between our cognitions and the perceived world, transforms what is really a psychological peculiarity into a general truth. This possible subjective peculiarity may also explain why Whitehead sometimes speaks of an eternal object, which when taken by itself is a bare class concept, as if it actually had some kind of shadowy existence.

VII

As far as the realm of eternal objects is concerned, Whitehead's theory might also be said to have some connection with applications of the mathematical discipline of combinatorial analysis. As we have noted, Whitehead is applying a system of logical structures to the physical world and psychological experience. These structures are related together in the form of a hierarchical series of grades or levels.[1]

The analytical character of the realm of eternal objects may under an approximation be thought of as the division (or Boolean expansion) of a universe class into a series of levels. On the first division it divides into two sub-classes of $n-1$ level; on the second, into four of $n-2$ level; on the third, eight of $n-3$ level; and so on. The number of sub-classes on each level can be represented by an increasing power of two. The whole division scheme may be

[1] Weyl's remark on Brouwer's conception of the continuum as an 'extensive whole' and its combinatorial division are of interest here. cf. H. Weyl, *Philosophy of Mathematics and Natural Science*, pp. 52-3, Princeton, 1946.

mapped out topologically, giving us a structure rather like a genealogical tree; thus,

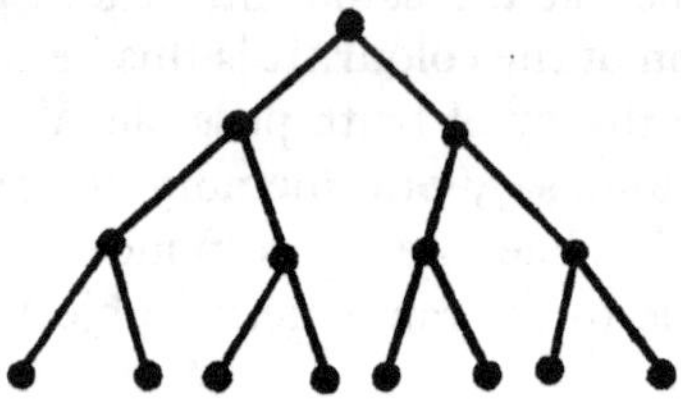

Whitehead's realm of eternal objects would also seem to bear some resemblance to the ramified theory of types, read in the direction of the highest to the zero type. In *Principia Mathematica* the construction of the hierarchy of types starts with first-order functions whose arguments are all individuals, continues with second-order functions having first-order functions as arguments; then with third-order functions having second-order functions as arguments. We can thus proceed indefinitely. So that if the highest order of variable occurring in the function is a function of the nth order, then the function in which it occurs is of the $n + 1$th order.[1]

Weyl, who constructs the theory of types in terms of an axiomatic system U, tells us[2] the theory can easily be represented by means of a topological tree; we need merely to indicate the highest level by a point, the types of a lower level by further points, and join them together by lines or branches. Each type corresponds to a complex eternal object in Whitehead's system; any type may be indicated by a subtree, whose branches end with the ground type. Similarly, each complex eternal object defines an abstractive hierarchy.

We think that the analogy we have drawn between these logico-mathematical notions and Whitehead's doctrine of abstractive hierarchies is not entirely unjustified. When Whitehead discusses abstractive hierarchies, he is not primarily concerned with collec-

[1] Cf. Russell and Whitehead, *Principia Mathematica*, Vol. 1, pp. 53-6 Cambridge, 1925. Though the theory of types was mainly Russell's preserve, Whitehead also showed an interest in types and systematic ambiguity. Cf. B. Russell, 'Whitehead and *Principia Mathematica*', *Mind*, Vol. LVII, No. 226, p. 138.

[2] 'Mathematics and Logic', *The American Mathematical Monthly*, Vol. 53, 1946, pp. 6-8.

tions of universals such as red, blue or green, but rather, 'with the investigation of patterns of connectedness in abstraction from the particular relata and the particular modes of connection'.[1] We 'are in the realm of complete and absolute abstraction'.[2]

SUMMARY

In his discussion of the realm of eternal objects, Whitehead is primarily concerned with extensive relationships in abstraction from any particular sensory structure. An eternal object has (1) an individual essence, i.e. is a colour, shape, etc., and (2) a relational essence—its pattern of connection.

The relationships of eternal objects to each other are analysable into an indefinite number of limited structures. Thus we can have knowledge of a specific object though remaining ignorant of the characters of other entities to which it is related. On a more concrete level Whitehead would say that though, for example, nature is analysable into a multiplicity of finite structures (or objects) each of which have their own individuality, they are nevertheless related together in one system.

A complex eternal object having a set of simpler elements subsumed under it, is called an abstractive hierarchy and may be either finite or infinite. Our experience has associated with it an infinite abstractive hierarchy in the form of a spatio-temporal perspective, together with a number of finite ones occurring in our mental states. Whitehead assumes that the possibility of the correspondence theory of truth arises from the correspondence between the eternal objects exemplified in perception, and those appearing in memory and imagination. On a logical level the realm of eternal objects bears some resemblance to the hierarchy of types.

[1] A I, p. 197. [2] S M W, p. 27.

CHAPTER V

THE ALGEBRAIC METHOD

I

IT may be said that all our evidence to the effect that Whitehead models the framework of his philosophy upon logical principles (leaving for the moment physical factors in the background) is largely circumstantial, based upon attempts to set up correspondences between different parts of his writings. For the passages we have quoted, other passages might be cited which would be reasonable evidence for some other interpretation. The choice of the best fitting interpretation would then be largely a question of judgment, the appreciation of Whitehead's work as a whole, and the resultant degree of intelligibility acquired by his more obscure writings. But even if our interpretation survived these tests, philosophers might still continue to doubt its validity.

There is, however, a piece of direct evidence to hand—the missing link, as it were, in our chain. Unlike most of Whitehead's later writings it is unambiguous, telling us precisely what sort of method he is adopting, for better or for worse in his philosophy. He is answering John Dewey, who has asked him to decide between the 'genetic functional' interpretation of first principles, and the 'mathematical formal' interpretation. Whitehead, however, declines to make the decision, as he thinks they both have their place in the Philosophy of Organism.[1]

'Our present problem,' he tells us, 'is the fusion of the two interpretations. The historic process of the world, which requires the genetic-functional interpretation, also requires for its understanding some insight into those ultimate principles of existence which express the necessary connections within the flux.'[2] In other words, the events which need to be interpreted in terms

[1] 'Remarks', *Philosophical Review*, 1937, pp. 178-81. Reprinted in *Essays in Science and Philosophy* as chapter 'Analysis of Meaning'.

[2] Ibid, p. 179.

of process also require for their understanding some insight into those principles which connect them together in our experience.

More light is thrown on this point when he explains exactly what he means by 'metaphysical necessity'. The clearest example of this is to be found, according to him, in 'the inter-relations of the specific multiplicities of groups of individual things' which issue 'in meaningful relations amid the accidents of history'.[1] Apparently, we are best able to study metaphysical necessity, which he identifies with the relations between events, by the algebraic method. The presupposition underlying such a view is that these experienced connectives are reducible to logical relations which can then be treated algebraically.

Whitehead describes the bare essentials of this method: it proceeds by selecting a few notions of the simplest inter-connections of things, i.e. types of relation. The fundamental assumption is that these basic connectives retain an invariable meaning throughout the algebraic development of patterns, i.e. that these relations retain their identity. By means of the device of the real variable, such symbols as p, q, r, x, y, z, u, v, w, are used under the assumption that each symbol in its repetitions throughout the complex pattern, indicates one and the same individual thing.[2] In the particular case considered by Whitehead, the basic connectives represent extensive relations, whilst the 'real variables' indicate the related events or accidental factors. Hence the system of events could be represented in the form $R(x \,.\, y \,.\, z \ldots n)$; and in this way we would arrive at its logical structure.

And this is what Whitehead appears to be doing in his account. He applies the algebraic method to the field of direct experience in the same way as in M C he applied it to the notion of the physical world made up of lines of force. Thus he can state 'These principles of algebraic symbolism express the concurrence of mathematical formal principles with accidental factors. This concurrence is inevitable for the production of meaningful composition.'[3] In other words, they express the 'manner' in which the relationships within facts relate events into meaningful patterns—the trees, houses and animals of our everyday existence. To make it clear beyond doubt that he is applying these principles to the 'relations amid the accidents of history', he asserts 'The basic connectives

[1] Ibid, p. 179.
[2] Ibid, cf. p. 183.
[3] Ibid, cf. pp. 183-4.

are the relevant mathematical-formal principles. The real variables are the unspecified accidental factors.'[1]

Whitehead then substitutes logical relations for the observed connectives, and 'real variables' for the events or accidental factors, which are thus toned down to refer to any entity which can be related in this manner. By such means we are enabled to disengage the relations within fact (the structure of our experience) and to express them in logical form.

As the above account deals with entities arrived at as a result of a process of logical substitution, and is not therefore a description of direct experience, Whitehead qualifies his previous statement. 'But the connection of the accidents is not a mere mathematical-formal principle. It is the concrete accidental fact of these accidents as thus connected.' This, he tells us, explains the vagueness which shrouds our metaphysical insight. 'We are unable to complete the approximation of disengaging the principles from the accidents of their exemplifications.'[2]

Whitehead recognizes that we live in a contingent world full of variable detail; that in our direct experience we are not aware of such abstract clean-cut logical relations, which are only disentangled in contemplative thought. These relations are suffused by concrete events, exhibiting the beauty and the colour which the poets felt had been left out of nature by the abstractions of the scientist. As Whitehead points out when speaking of the poetry of Wordsworth, Wordsworth's consistent theme is that the important facts of nature elude the scientific method: 'That is why he laughs with the daffodils, and finds in the primrose "thoughts too deep for tears" '.[3] The nature poetry of the romantic revival was a protest 'against the exclusion of value from the essence of matter of fact'.[4]

What Whitehead is trying to do is to unite into one philosophic outlook the aesthetic intuitions of mankind, and the formal principles discovered in nature by scientific analysis, thereby giving these abstract principles a qualitative content or value. In this case the two distinct senses of value (the mathematical and the aesthetic) have the same referent, since the extra-logical values assigned to this scheme are particular experiences. Whitehead considers that aesthetic and other tertiary qualities are as directly given in our

[1] Ibid, p. 184. [2] Ibid, p. 184.

[3] S M W, p. 104. This seems to be a misquotation from the *Ode on Intimations of Immortality*.

[4] S M W, p. 118.

experience of nature as are the observed reds, blues and greens, usually termed sense-data.

We thus return to the point made above. 'The connection of the accidents is not a mere mathematical formal principle.' In his description of nature Whitehead is not dealing with an abstract set of relations, but with events having a definite perceptual content. The perceived events (or the historic process of the world) are the definite values which suffuse these logical relations. And this, he tells us, explains the difficulty we have in formulating metaphysical first principles. 'Our metaphysical notions are an approximation. They represent such disengagement of necessity from accident as we are able to attain.'[1] It is interesting to see that the sort of metaphysical notions Whitehead has in mind here are more like the concepts found in *Principia Mathematica* than in the *Logic* of Hegel.

Whitehead now makes what is perhaps the most significant remark of his whole paper, as it puts in a nutshell his philosophic method. He asserts that 'the algebraic method is our best approach to the expression of necessity, by reason of its reduction of accident to the ghost-like character of the real variable'.[2] From this it appears that the best method we have on our hands for the expression of metaphysical first principles, 'the necessary connections within the flux', is the algebraic. By its means we can disengage these principles from the historic process of the world, replacing the accidental factors, i.e. events, by 'real variables'. In this manner we obtain a pattern of logical relationships which, owing to its complete generality, refers to unspecified events, past, present or future. We thus arrive at a key notion of Whitehead's philosophic scheme, the extensive continuum or the logical structure of fact.

The reader may perhaps still doubt our assertion that the method Whitehead urges us to adopt in philosophy resembles that of symbolic logic, which he equates with mathematics in its most general sense, i.e. the investigation of abstract patterns of connection. We may therefore also note Whitehead's statement that symbolic logic bears a relationship to metaphysics, since 'the reformation of Logic has an essential reference to Metaphysics. For Logic prescribes the shapes of metaphysical thought.'[3]

Whitehead concludes this part of his discussion by asserting that our mathematics and our symbolic logic, as hitherto developed,

[1] 'Remarks', p. 180. [2] Ibid, p. 184.

[3] Foreword to W. V. Quine's *System of Logistic*, pp. ix-x.

represent only a minute fragment of the possibilities of the algebraic method. He believes that, owing to the complete generality of this method (with its emphasis on pattern rather than number or quantity), it can be applied to fields which so far have not been amenable to mathematical treatment. After quoting from the *Sophist* and suggesting that Leibniz was sympathetic to such a view, no doubt because he attempted to apply the analytic method of mathematics to metaphysics, Whitehead concludes by saying, 'And now having evoked such support, I can cease the defence of the attempt to bring together the genetic-functional and the mathematical-formal methods in one philosophic outlook'.[1]

II

We are now brought to Whitehead's statement which has won notoriety for itself among American philosophers, though it seems to have passed relatively unnoticed in this country. 'We must end', he tells us, 'with my first love—Symbolic Logic. When in the distant future the subject has expanded, so as to examine patterns depending on connections other than those of space, number, and quantity—when this expansion has occurred, I suggest that Symbolic Logic, that is to say, the symbolic examination of pattern with the use of real variables, will become the foundation of aesthetics. From that stage it will proceed to conquer ethics and theology.'[2]

It is quite clear then that part of Whitehead's programme is to tackle ethical as well as theological problems by means of symbolic logic. With this he is at one with Leibniz's 'universal language'. Leibniz believed that with such a medium 'we could reason in metaphysics and in ethics very much as we do in geometry and analytics', and it is very likely that he would also have included theology. The above passage from Whitehead would seem to controvert Lowe's statement[3] that, unlike Leibniz, Whitehead never considered Logic as the essence of philosophy, if by this Lowe has in mind philosophic method. Indeed, Whitehead seems to have gone much further than Leibniz on this point. Owing to the generality of modern symbolic logic, he patterns his metaphysics on the 'axiomatic method'. Lowe does not appear to have distin-

[1] 'Remarks', p. 184. [2] Ibid, p. 186.

[3] 'Development of Whitehead's Philosophy', in the Schilpp volume, p. 19.

guished between a technique used in investigating experience and the content of that experience itself (i.e. between a deductive system and its field of application). As mathematicians, both Leibniz and Whitehead clearly recognized this distinction in principle, though there is a tendency in their philosophical writings to blur it, and to transform logical into existential principles.

No wonder Urban, when commenting on the above passage, asserts that it is as enlightening as it is staggering. He finds it staggering because to say that symbolic logic, however developed, can conquer aesthetics, to say nothing of ethics and theology, contradicts his previous conception of Whitehead's philosophy, namely, that it found a place for the qualities and values in nature to which the poet is so sensitive, and which slipped through the meshes of the logical net of the scientist. Urban regards it as enlightening, because it confirms what he has long suspected, that the tacit assumption throughout Whitehead's entire philosophy is really the primacy of science and mathematical logic.[1]

If, however, we translate the term 'aesthetic' to mean simply direct experience which has tertiary qualities such as beauty, as well as secondary qualities, and use 'value' in its duality of senses, this apparent contradiction does not, we think, arise. The symbolic examination of pattern with the use of real variables then becomes the foundation of our methodological approach to the study of direct experience. Light is also thrown upon the apparent dualism of PR; the patterns of relations are the necessary permanent elements, the real variables representing the accidental or fluent factors.

As we have seen, what Whitehead seems to be doing in the last chapter of PR is applying symbolic logic to natural theology. Consider, for example, his statement that 'God's conceptual realization is nonsense if thought of under the guise of a barren, eternal hypothesis', as it performs 'an efficacious rôle in multiple unifications of the universe'.[2] He is not talking here about some ethereal spirit casting its spell on the universe. He is rather asserting that the pattern of relationships pervading our experience is not merely a formal one, but is to be thought of as connecting concrete accidental facts. We may regret, however, that he could not find a less ambiguous way of expressing this truth.

According to Whitehead, if we abstract the logical (static) and factual (fluent) elements in experience, and give them each an

[1] Urban: Schilpp, cf. p. 310; cf. also p. 321. [2] PR, cf. p. 494.

independent existence, then their interplay involves contradiction at every step. In this way we obtain the antitheses of the last chapter of P R.

The upshot of our discussion is that Whitehead is using the algebraic method in his philosophy as a technique for investigating direct experience. For example, speculative philosophy as defined by him in Chapter I of P R has the character of a formal necessary system of ideas which are generalizations of certain basic features in experience. This seems to be in keeping with the main thread of thought running through his writings from *Universal Algebra* onwards.

The resultant analysis of experience into structural and qualitative elements gives rise to his whole doctrine of permanence or repetition of character amid the flux of events. There is, of course, empirical support for this doctrine. We cannot help but be aware that nature exhibits certain broad phases of order. (That they are, as he claims, of 'social' origin, due to the inter-relationship of events is, of course, only one possible explanation of the phenomenon.) Though the logical scheme Whitehead substitutes for this intuition might be a convenient tool for investigating nature, it should not, however, be transformed into a constituent of direct experience. Hence, it is questionable whether Whitehead's account of 'metaphysical necessity', or 'the necessary connections within the flux', refers to anything more than this conceptual superstructure.

What should be noted is that Whitehead advocates the use of symbolic logic in such qualitative fields as ethics and theology. But he only gives us general suggestions as to procedure, and does not work them out in any detail. For example, his account of God and the World in P R merely represents them as being related to each other, in the way a logical function is related to its values.

We do not deny that as a technique for dealing with fields not readily amenable to quantitative methods, in such sciences as sociology, psychology and biology, there may be great possibilities for applied mathematical logic. Indeed Woodger[1] has attempted something of the sort in biology and Hull in psychology, and there have also been some attempts in other fields, especially cybernetics. Whitehead, however, seems to have been a pioneer here. In M C he applied this method to the material physical world, and in P R to experience in general, except that in the latter work he expressed his ideas in extremely obscure language. His treatment in P R is

[1] Cf. *The Axiomatic Method in Biology.*

consequently far from rigorous, as clearly defined symbols have been sacrificed for ambiguous verbal phrases. It may, of course, be the case that large ranges of experience defeat precise symbolism.

One can draw an interesting comparison with Leibniz. The logical argument that the predicate is contained in the subject led Leibniz to the denial of interaction between substances, and to his theory of monads, the monads being modelled on the subject-predicate form of proposition—as independent substances with inhering qualities. In a similar way Whitehead's view appears to be based upon an account of the world expressed in terms of symbolic logic. In this account emphasis is put on relations rather than upon the specific or generic character of the terms. We may then say that the logic of *n*'adic relations forms the model for Whitehead's view. It seems, at least on its logical side, to have led to his doctrine of 'immanence', according to which the world is made up of an infinite number of atomic entities, which are related to each other as terms within an *n*'adic logical function. We may therefore argue that just as Leibniz's metaphysics was based upon the subject-predicate pattern, or the *S-P* logic of his day, so Whitehead's view is modelled upon the polyadic propositional function, or modern mathematical logic.

As Whitehead conceived the relations between events to be serial in character, he thought of this set of relations in the form of a continuum, underlying the world, past, present and future. His whole view of 'societies' (e.g. physical objects) can be reduced to a set of serial relations, having a concrete interpretation. The particular electro-magnetic (or experiential) characters are the terms which form the fields of these relations.

There seems no reason why the more formal parts of P R should not be expressed in symbolic terms. Such a treatment would be possible as his account contains logical as well as empirical factors. We could, for example, symbolize the universe of events in the form of such a relational system made up of subsidiary patterns (or physical objects), each of which is further sub-divisible. Whitehead seems to have been sketching such a method of analysis in his chapter on 'Abstraction' in S M W.

The method of symbolic logic (the mathematical formal principle) and his conception of electro-magnetic or experiential (i.e. genetic functional) characters form the two key notions of Whitehead's philosophy.

SUMMARY

Is there any direct evidence that Whitehead models the framework of his philosophy on logical principles? In reply to Dewey, Whitehead tells us that the historic process of the world requires for its understanding some insight into the necessary connections within nature and this, he claims, can best be studied by the algebraic method. What Whitehead's suggestion comes to is that we can obtain an understanding of 'metaphysical necessity' by substituting logical relations for the observed connectives, and 'real variables' for the related events or accidental factors.

Whitehead tries to unite in one philosophical outlook the aesthetic intuitions of mankind and the formal principles discovered in nature by scientific analysis. One might say that just as Leibniz's metaphysics was based upon the subject-predicate pattern, or the *S-P* logic of his day, so Whitehead's view is modelled on mathematical logic with its emphasis on multi-relational systems.

CHAPTER VI

THE EXTENSIVE CONTINUUM

WHEN Whitehead speaks of the extensive continuum, it is clear that he is simply referring to a system of logical relations. In that case, we may ask, how can it be a constituent of actual fact, as he claims? It is almost a truism that logical (or mathematical) entities and contingent facts have diverse properties, and should not therefore be confused if one wishes to avoid paradoxes; unless he means that he is dealing with a system of entities which are definitions (or explications) based upon certain abstract features of experience. But then these definitions, unlike statements about the original phenomena for which they have been substituted, will only need the test of consistency. 'Thus no external verification of definitions is required in mathematics, as long as it is considered merely as mathematics.'[1] Whitehead, as we shall see, goes beyond this limited sphere and tends to give these notions an existential import.

Consider, for example, his account of the extensive continuum as one relational complex which underlies the whole world past, present and future.[2] But can such a complex of abstract relationships underlie the empirical world in this fashion? This is on a par with supposing that the parallels of latitude and the meridians of longitude, which merely refer to certain abstract features of the earth, underlie it in the form of a framework. Whitehead, as we have seen, is really dealing with an abstract logical system, namely a set of relations ordering a certain group of entities which thereby obtain a definite structure. However, in our experience we do not come across anything corresponding to such a complex (or multitermed) logical system which can take events as values. Further, it is doubtful whether there is an empirical class of things—events—to be thus related. The notion of an event only seems to be arrived at as the result of a process of rational reconstruction.

Take as another example his statement, 'The notion of a "continuum" involves both the property of indefinite divisibility

[1] U A, p. vi. [2] P R, cf. p. 91.

and the property of unbounded extension. There are always entities beyond entities.'[1] Now we do not object to this if he is talking about the properties of a set of related entities making up, say, the mathematical continuum stretching backwards and forwards to plus and minus infinity. But this remark may certainly be queried if it is applied to actual experience. Indefinite divisibility and unbounded extension do not seem to be properties of the empirical world. At the most on Whitehead's view they could only be said to be schematic programmes based on certain relations discriminated by us in perception—for example, that various parts of our perceptual field are connected together and that each part includes smaller parts. The information that one can split things up indefinitely, and that there are always 'entities beyond entities', can only be arrived at by inspecting the abstract model.

Whitehead is then not merely supposing that there will be an order, but also that there will be an infinitude of events to have this order. Further, as at the present moment the empirical world and the infinite scheme have not the same number of terms, can one really say they have the same structure? There seems little evidence to show that the world is infinite in space and time (or is an infinite system of events). In fact, what evidence there is points the other way.

Though the meaning of an event is closely tied up with the 'before and after' relation, what reason has one for supposing that the aggregate of future events will have such a serial order?[2] The relations between them may not be of a serial type, but may be symmetrical. After the next event we may become omniscient and take in the whole past and future at one sweep; or it may be a cyclical structure—after a certain number of events the past events may recur. Hence, the aggregate of future events may have a variety of possible structures, depending on the relations by which we choose to connect them. The notion of an event would thereby take on different meanings with different relations.

However, all that Whitehead might mean when he assumes that the present is systematically related to the future, is that future events must have some kind of structure, i.e. be related to each other in some way as yet unspecified. But he expressly tells us that the extensive continuum is a complex of entities united by the

[1] P R, p. 91.

[2] The classical mathematical notion of continuity applies only to series which have the property of being asymmetrical, transitive and connected.

various allied relationships of whole to part, or of overlapping and contact, and other derivative relationships.[1] The notion of extensive connection is then already a specific type of relation and hence specifies the structure of future events, information not given in our present experience. Whitehead might, however, argue that one cannot sensibly talk about the future unless one accepts, as a minimum assumption, some such structure. It is, for example, meaningless to talk about future events without postulating the 'before and after' relation. He would also probably argue that the relation of 'extensive connection' is basic to any other extensive relationship.

The fundamental assumption underlying Whitehead's argument seems to be that in some way logical and mathematical entities (either as abstractions from experience or as postulates) can give us an insight into the future. As Bridgman points out in another context, the implication is that mathematics is not limited simply to the results that have been investigated, but goes further and says something about anything we might do at any time in the future. To do this, Bridgman concludes, is simply to confuse logical with physical possibility.[2]

Whitehead believes that 'The extensive continuum is "real", because it expresses a fact derived from the actual world and concerning the contemporary actual world'.[3] One might reply that, even if the continuum is to be thought of as a derivative abstraction from certain recurrent features of our experience, it is certainly not real in the same way as are the particular facts of the recurrence of night and day, the seasons and the rhythmic biological processes. We already deal here with a piece of mathematics having more general properties than the actual empirical processes; with relations and classes of relations which have been substituted for the crude perceptual data. Whitehead nowhere makes it clear how entities which he seems to regard as ideal limits can have a real existence—how, for example, 'The reality of the future is bound up with the reality of this continuum'.[4] As the future is not yet given, where are these relations to find a home for themselves?

The extensive scheme would seem to be a piece of applied mathematics in the sense given to that term by Whitehead. In one sense, he tells us, there is no applied mathematics. 'When once the fixed conditions which any hypothetical group of entities are to

[1] P R, cf. p. 91.
[2] *Nature of Physical Theory*, cf. pp. 52-3.
[3] P R, p. 92.
[4] P R, p. 92.

satisfy have been precisely formulated, the deduction of the further propositions, which also will hold respecting them, can proceed in complete independence of the question as to whether or no any such group of entities can be found in the world of phenomena.'[1] And he cites as examples rational mechanics and the theory of perfect fluids.

If our interpretation is correct, Whitehead in his discussion of the extensive continuum is really concerned with a logical method of representing nature. The necessity which he assumes to be found in nature is rather a story about the logical connections of the abstract entities constituting the scheme. Though Whitehead may perhaps at times give one the impression that he thinks otherwise, he is not giving a description of fact in the same way as the biologist describes the circulation of the blood. When he asserts, for example, that 'Actual entities atomize the continuum',[2] this is certainly not descriptive of a perceptual process. Even supposing we are conscious of a passage of events, we are certainly not aware that it weaves itself upon a background of logical relations. The process he is talking about seems really to be a schematization of certain aspects of experience—a piece of applied logic, in fact. What Whitehead seems to be doing is setting up a logical scheme symbolizing, for example, the perceived passage of nature, in the same way as in mathematical physics we might symbolize the passage of the planets round the sun. However, this does not justify us in assuming that the formulae are logical pictures of the actual occurrences. The whole notion of events as atomizing the extensive scheme, or the giving of empirical content to logical structures, arises from the assumption that logic can serve as a model for direct experience.

Finally, we need to say a few words about the notion of the continuum itself. It has often been pointed out that there is a radical difference between the continuity of our experience and mathematical continuity which applies only to series. By experiential continuity we refer to the fact that the perceptual field is given as a connected whole which cannot be split up into ultimate simple elements. Mathematical continuity, on the other hand, deals precisely with such elements, i.e. an infinite collection of individuals arranged in a certain order.

It is clear from Whitehead's conception of the extensive continuum as exhibiting the properties of the inclusion, overlap and contact of regions, that he has not the classical mathematical

[1] *Essays in Science and Philosophy*, p. 203. [2] P R, p. 92.

continuum in mind. It is in any case difficult to see how he could conceive it as a closely packed infinite series of points, since he regards points as derivative from these extensive relationships. In this respect his position would seem in some ways to resemble the mathematical intuitionist's notion of the continuum, where the fundamental relationship is taken to be that of part and whole. In the mathematical continuum of Brouwer, for example, the individual real number is defined as 'an *infinite sequence of nested division intervals of increasing level*'.[1]

SUMMARY

Whitehead's whole account of events as atomizing the extensive continuum, or the giving of content to logical structures, arises from the assumption that logic can serve as a model for direct experience. We enquire how, if the extensive continuum is simply a system of logical relations, it can be, as he claims, a constituent of actual fact. It is argued that, since logical entities and contingent facts have diverse properties, they ought not to be confused.

We note the difference between the mathematical continuum and the continuity of experience. Whitehead's extensive continuum resembles in some respects the intuitionist's continuum, which is also based on the part and whole relationship.

NOTE

Leibniz and Whitehead on Space and Time

IT is interesting to compare the views of Leibniz and Whitehead on space and time. Taking Leibniz first, we see that for him space and time were not something actual, but merely abstracted by the mind from the things observed. Space, for example, is nothing but an order of possible situations, time an order of possibilities which are successive. Thus he is led to say 'Space and time taken together constitute the order of the possibilities of a whole universe, so that these orders (that is space and time) square not only with what actually exists but also with whatever might be put in its place, as numbers are indifferent to whatever can be *res numerata*'.[2]

[1] H. Weyl, *Philosophy of Mathematics and Natural Science*, pp. 51-53.

[2] 'Réplique aux Réflexions de Bayle', quoted from Latta's ed. of the *Monadology*, p. 102.

It would appear that the more general notion underlying Leibniz's view is that of order, space and time being two varieties of it. For example, when we deal with space we are not concerned with the entities situated therein, but merely with this ideal series of positions. Similarly, in the case of time we are not concerned with any specific entities, but merely with their order of succession. As we have seen, Leibniz was influenced in his view that space and time were orders of positions and instants by the order of numbers and their generality. He clearly saw that points and instants are in this respect similar to numbers; they are indifferent to whatever is situated in them, for owing to their ideal nature they are perfectly general.[1] And this was undoubtedly one of his reasons for rejecting Newton's notion of absolute space as an existent entity.

The resemblance to Whitehead's views may not at first sight be very apparent. If, however, we were to combine the spatial and temporal orders so that they formed a framework in which different entities could appear, we should approximate to Whitehead's four-dimensional spatio-temporal manifold made up of a series of event-particles. If we were to go a step further and concentrate upon the notion of order, the resultant framework would come very close to his concept of the extensive continuum.[2]

There is some ground for believing that Leibniz's notion of a pre-established harmony is not so very far removed from Whitehead's concept of the extensive continuum. According to Leibniz, all the monads in the universe, though independent of each other, are so ordered that certain mental events can follow upon certain physical events without there being any causal connection between them. If we substitute events for the monads, and the extensive continuum which orders these events for the pre-established harmony, the resemblance to Whitehead's views will readily be seen. The main difference between Leibniz and Whitehead is that the latter regards this abstract order as ultimately arising from the causal relations between events, while Leibniz takes this order as being divinely pre-established and imposed upon his monads.

[1] Cf. 'Correspondence with Clarke', *Philosophical Writings: Liebniz*, ed. Morris, p. 215. 'The parts of time and of place, taken in themselves are ideal things; thus they are perfectly alike, like two abstract units. But this is not the case with two concrete unities, nor with two actual times, nor with two occupied, that is to say truly actual, spaces.'

[2] We have attempted to show that this notion is identical with Whitehead's Primordial Nature of God.

CHAPTER VII

THE METHOD OF EXTENSIVE ABSTRACTION

I

AS there have been a number of accounts of the Method of Extensive Abstraction, it will be more useful for our purpose if we merely summarize it and then examine some of the criticisms which have been brought against it. One of the factors which apparently led Whitehead to develop it was the rejection by the theory of relativity of the conception of space as an aggregate of absolute positions. On the relational theory of space, geometry is the science of the complex relations existing between physical things. The method, which is applied to time as well as to space, has for its aim definitions of point, line, instant, etc., as logical functions of what is given in sense.[1]

Whitehead pictures the world as made up of events having volume and duration and having to each other whole and part relations. The key notion used by the method is that of convergence to simplicity with diminution of extent. We start with some large enough event and analyse it into a convergent series of successively smaller events. Such a route of convergence is called an abstractive set or class. It fulfils the following conditions: (i) of any two of its members one extends over the other,[2] and (ii) there is no event which is extended over by every event of the set. Each set is thus composed of an infinite series of successively smaller events, which converge without ever arriving at a terminal event. Every point, line, surface and instantaneous volume is either a class of events or a class of a higher order founded on events. In this process of convergence certain relations will be simplified or excluded, and in this way we arrive at the ideal simplicity of nature at an instant.

The method has, however, been criticized on the ground that geometrical elements, namely points, lines, volumes and areas, are

[1] For an account of the method, cf. P N K, Part III, and C N, Chapter IV.

[2] This might be called the 'condition of inclusion'.

presupposed in this definition of an abstractive set. In answer to this, Stebbing[1] asserted that, since these elements are really defined in terms of convergent sets, the definition is not circular. We are, she argues, sensibly aware of series of volumes which have relations similar to the convergent sets which define a point or a plane, and we are also aware of the difference between these two types of convergence. Our only difficulty is to state in formal logical terms these differences which we can all vaguely see and feel. Nevertheless, it can be argued (*a*) that this still does not get over the difficulty that we never perceive the smaller events of the convergent series; and (*b*) that this method presupposes some sort of conceptual extrapolation to deal with the 'condition of inclusion', which presupposes the notion of an infinite series.

Whitehead's account of the method in P R[2] differs somewhat from his earlier view. Instead of dealing with events having a whole and part relation to each other, he deals with abstract regions having the relationship of extensive connection. Such relations are regarded as being more fundamental in character than the whole and part relation. Two regions *A* and *B* are said to be extensively connected if there is no region separating them. They are mediately connected if they are both connected with a third region. Connection and mediate connection are both symmetrical, non-transitive relations. The whole and part relation is now defined in terms of extensive connection. Thus region *A* is said to include region *B* when every region connected with *B* is also connected with *A* (i.e. *A* is part of *B*).[3] As these regions are completely general, he tells us that we can see, by reference to the particular case of three-dimensional space, i.e. by the use of spatial diagrams, that the various types of abstractive sets can converge to different kinds of geometrical elements. But he warns us that such diagrams may be misleading. They introduce irrelevant spatial features which have to be defined in terms of the fundamental notion of extensive connection.

[1] L. S. Stebbing, *A Modern Introduction to Logic*, 5th edition, pp. 450-1.

[2] P R, pp. 416-26.

[3] A further condition is introduced in P R, that of non-tangential inclusion. This condition excludes any two members of the abstractive set from having a common boundary, i.e. one must be always included in the other without touching. It ensures that the ever smaller regions of the set do not approach the boundary of the set and thus allows the set to converge to a unique position. (P R, p. 420. Definition 9.)

II

Whitehead's Method of Extensive Abstraction has recently been severely criticized by Grünbaum.[1] He starts from Broad's account,[2] which asserts that Whitehead tries to exhibit geometrical notions as derivative from the data given in sense. Grünbaum argues that, to implement this programme, points must satisfy two conditions: (1) they must have to each other the kinds of relations which geometry demands, (2) they must be related to finite areas and volumes in such a way that the latter can be exhaustively analysed into sets of points. Among other things, he argues, (*a*) that the method is vitiated by Zeno's mathematical paradox of plurality, because it violates these conditions, (*b*) that the convergence of the abstractive classes is fatally ambiguous, and (*c*) that the abstractive classes do not belong to the domain of sense-awareness.[3]

To start with (*a*); one of Zeno's paradoxes endeavoured to show that, since spatial and temporal intervals are extended, it is self-contradictory to regard them as aggregates of unextended points and instants. Grünbaum contends that part of contemporary mathematics rests on this conception that an interval is literally composed of an infinite number of unextended point-elements. However, since every denumerable point has zero measure, an interval, he argues, can be consistently regarded as an aggregate of points only if this aggregate is super-denumerably infinite. In other words, it has to be regarded as belonging to a higher class than the points which compose it. Any epistemological reconstruction of modern geometry and physics must therefore preserve the super-denumerability of intervals which guarantee continuity, since the theoretical measurables of physics range over a mathematical continuous domain. Whitehead is then faced with the problem of discovering in sensed nature a corresponding super-denumerable infinity of abstractive sets which will define these points. As such an attempt is doomed to failure, Grünbaum concludes that the method breaks down on this score alone.

In the case of (*b*) Grünbaum points out that Dedekind's definition of irrationals clearly differentiates between every two irrationals, however close together these may be in magnitude, as otherwise the definition would be rejected as fatally ambiguous.

[1] 'Whitehead's Method of Extensive Abstraction'. Adolf Grünbaum, *British Journal for the Philosophy of Science,* Nov. 1953, pp. 215-226.

[2] *Scientific Thought,* cf. pp. 38-52.

[3] Ibid, p. 216.

He asks us to consider two distinct neighbouring points such as $x = 0$ and $x = 10^{-1000}$,[1] which differ in identity so much that they are separated from each other by a continuum of points. But, in virtue of the infinite divisibility of intervals, we cannot, he argues, know from sense-perception that there exist two different abstractive classes defining these points. As the method does not therefore enable us to differentiate between the diverse points in systems of mathematical physics, it obliterates, according to Grünbaum, 'the very precision of meaning made possible in the statement of physical laws by the use of real variables'.[2]

III

Grünbaum's criticism of the Method of Extensive Abstraction stems to a considerable extent from his acceptance of Broad's account, which assumes that the method is based on sensationalist foundations. Part of the misunderstanding no doubt arises from the fact that Whitehead does not always make it sufficiently clear whether his method is to be taken as an algorithm or as an exact description of some actual process of convergence.

According to Russell, Whitehead was examining from the point of view of mathematical logic how we may define in terms of empirical data the entities that traditional geometry considers as primitive. However, the method, as Russell points out when he compares it with Nicod's attempt to construct a geometry from sense-experience, starts from knowledge of the completed mathematical system which is the object to be attained, and goes back to entities more analogous to those of sense-perception. Nicod's method, on the other hand, he says, follows the reverse order; proceeding from the data of perception, it tries to attain the various geometries which can be built up out of them.[3]

Nicod himself suggested that Whitehead's contribution could be

[1] Ibid. cf. p. 219. [2] Ibid. p. 219.

[3] Preface to Nicod's *Foundations of Geometry and Induction*, cf. p. 7. What Nicod does is to take as his starting point a simplified form of sensation. He postulates an animal possessing only the sense of hearing and a perception of temporal succession, which produces notes of varying pitch as it proceeds up and down the keyboard of a piano. From this Nicod proceeds to postulate another hypothetical animal coming closer to man in his perceptions. It seems doubtful whether this method of constructing models of sense-experience really brings us any closer to our actual perceptions.

taken as the construction of a pure geometry rather than as an analysis of the real world. He contends that, since Whitehead's construction starts from an analysis of the terms and relations that nature presents, which already possess the properties of geometrical volumes, the method's axioms and theorems become an investigation of pure geometry—the geometry of volumes. The function of the method is to form a series of volumes which obey the laws of points, so that for every point in the geometry of points we can substitute the limit of a unit aggregate of volumes and *vice-versa*. For Nicod, then, Whitehead's whole account only applies to mathematical volumes for which (unlike points) a specific interpretation exists in sense-perception, provided we start from a sufficiently large size of volume.[1]

If Whitehead's method is to be taken as a piece of applied logic, as an abstract model, enabling us to get from volumes to points, the question arises how far the regions[2] of his system apply to the sensed volumes. Whitehead was clearly aware that the abstractive sets of convergent events could not get us to a terminal event. In C N he introduced the notion of a quantitative series of measurements, symbolizing the quantitative relations present in events, which he used as a model for the convergent sets of events. Though the quantitative series converges to a definite limit there is, however, no one-one correlation—the sets of events only approximate towards such an ideal simplicity.[3] Whitehead seems, however, sometimes to argue by analogy from the quantitative series to the corresponding convergent series of events, that if the latter has certain properties of the former, it will also have others which we can imagine. In this connection Russell made the interesting comment that, though the method is equally applicable to psychological and physical space, it does not in the former case yield continuity, unless we assume that sense-data which have a minimum size below which nothing is experienced always contain parts which are not sense-data. 'The full employment of Dr Whitehead's methods,' he concludes, 'therefore, belongs rather to physical space than to the space of experience.'[4]

[1] Ibid, pp. 40-43.

[2] Though it should be remembered that the volumes Whitehead starts from are undefined elements exhibiting certain quantitative characteristics and not necessarily those of geometry.

[3] Cf. C N, pp. 80-83.

[4] *Our Knowledge of the External World*, 1949 ed., p. 121.

Now though the system of volumes belonging to the mathematical model does converge to zero or a limit, Grünbaum would argue that this only applies to the spheres of co-ordinate geometry which are themselves first defined by the points of analytical geometry, which consequently begs the question. Further, ambiguity of convergence again occurs. We are unable to differentiate between two different neighbouring points,[1] since the abstractive classes corresponding to both alike converge to zero.

With regard to circularity, Whitehead might say (at least in P R) that the volumes he is concerned with are simple indivisible elements and thus do not presuppose points. And if the method is merely a technique for translating certain series of convergent volumes into points, the circularity criticism is not of moment, since this is a character of any axiom system. The postulate of 'continuity of inclusion', which determines that each event in the series covers and is covered by other events, would also require no justification, as the notion of an infinite series is taken over with the rest of modern mathematics—though in the light of some of his remarks in M T the notion of such a series may need reinterpretation.

As far as the criticism of ambiguity of convergence is concerned, a similar difficulty would arise in the case of the intuitionist continuum. According to Brouwer, there are pairs of real numbers a and b for which it is not known that either $a = b$ or $a \neq b$ or $a \# b$ ($a \# b$ being a stronger kind of inequality in which we can give a rational number separating a and b).[2] Since we cannot say on his view whether the infinite sets defined by such pairs of numbers overlap, the law of excluded middle breaks down. And this is what ambiguity of convergence seems to come to in such cases.

IV

Whitehead's position might be defended on the ground that we do at least in thought encounter the inexhaustibility of an abstractive set. Indeed, Whitehead himself has said that, if we assume the ideal infinite divisibility of space, this assumption makes a rather large draft on the theory of ideal perceptions, but that geometry vanishes unless it is made in some form.[3] Grünbaum would argue that this defence depends upon a tacit appeal to the infinite divisibility of

[1] Grünbaum, p. 220.

[2] Cf. S. C. Kleene, *Introduction to Metamathematics*, p. 53.

[5] A E, pp. 176-177.

space-time, defined in terms of relations between non-perceptual point elements. But how on such an interpretation, he asks, are we to interpret the axiom of infinity (which assumes there are infinite aggregates in the world) needed for the theory of real numbers and points?[1] Clearly, we do not perceive such aggregates, nor for that matter could we imagine them. However, apart from the rather dubious nature of the axiom, this criticism would only apply if one assumed that perceptual space (or the structure of appearance) was isomorphic with the real number continuum.

Whitehead is not entirely blameless. Though he recognizes in C N that one can only clarify the notion of a convergent series of events by means of a mathematical model, he seems at times to lose sight of this fact. For example, he appears to assume that we could by means of the method arrive at the serial order among moments in a time system which has the Cantor-Dedekind type of continuity.[2] Starting as he does from a system of whole and part relationships, it is difficult to see how he can arrive at this kind of continuity involving a non-denumerable infinity of points.

On the other hand, the confusions which Grünbaum finds in Whitehead's position partly arise from the former's assumption that physical space and time have necessarily the continuity of the real number continuum, the elements of which are hence non-denumerable. Whitehead himself has pointed out that we have no way of discriminating whether any kind of physical quantity (e.g. space) possesses the compactness of the series of rationals (which can have quasi-gaps in it) or the continuity of the series of real numbers. Though, he goes on, we assume the latter in calculations because of its mathematical simplicity, this assumption is unsupported by *a-priori* or experimental grounds.[3]

Grünbaum's criticism that the method obliterates the precision of meaning made possible by the use of real variables arises from just this belief. Brouwer, however, has argued that statements which concern all real numbers, i.e. all the values of a real variable, are to be interpreted in terms of the totality of natural numbers. He defines the individual place in the continuum not by a set but by a sequence of natural numbers created by free acts of choice. Whitehead's conception of the continuum as a connected system of regions rather than a series of discrete points brings him closer to the intuitionist's continuum, where the fundamental relation is that

[1] Grünbaum, cf. p. 226. [2] P N K, cf. p. 115.
[3] *Essays in Science and Philosophy*, cf. p. 203.

of part to whole rather than element to set. Further, it seems from Whitehead's assertion in M T[1] that every mathematical statement is about a process and its issue (e.g. $2 \times 3 = 6$) which holds of an infinite series as well as the multiplication table, that he emphasizes as the intuitionists do, the operational side of mathematics.

V

The ancestor of the Method of Extensive Abstraction is to be found in Whitehead's theory of interpoints (intersection points) by means of which he defined points in M C, and which is regarded there as a straightforward logical investigation,[2] epistemological considerations being excluded. This early work would appear to form the foundation for his later theory of events as well as for the method.

On this view, the points of space are complex relations defined in terms of classes of what he called 'linear objective reals'.[3] Linear objective reals he tells us, have properties which we associate with straight lines considered throughout their whole extent as simple indivisible entities—they are thus not to be conceived as made up of points. An ordinary geometrical line which has parts and segments is a class of points, and so is the ordinary plane. Whitehead attempts to specify the class of 'linear objective reals' in terms of lines of force. But he warns us not to identify the two, as the latter has ends, the former not. In C N we seem to get a similar position when Whitehead speaks of an absence of maximum and minimum durations in nature.[4] His concept of an ether of events in C N appears to be an extension of his earlier doctrine of objective reals. In both cases geometry is regarded as a study of a certain limited set of their properties. In M C he dealt with a certain polyadic relation R, the theorems of Euclidean geometry being an expression of certain properties of its field. In C N geometry becomes a characteristic of the structure of events.

A point in M C is defined in terms of a class of entities (objective reals) having a similarity of position. One notes the resemblance of this definition to the definition of number, for example, the cardinal number one, in terms of a class of similar classes, except that the relation he is dealing with is that of 'similarity of position'. The relation of interpoint order, where objective real a intersects

[1] M T, cf. p. 125.

[2] *Alfred North Whitehead: An Anthology*, cf. pp. 35-39.

[3] M C, cf. Part III (i).

[4] C N, cf. p. 59.

objective reals *a*, *b*, *c*, *d*, (the class of interpoints having a common position) at instant *t*, becomes in his later work the relation of extending over—each event extends over and is extended over by events. The notion of 'similarity of position', which has an ideal character, is now reached by a process of convergence. The bridge between his earlier and later work occurs in A E, where a point is defined as 'the class of extended objects which, in ordinary language, contain that point'.[1] This still has overtones of the definition of number in terms of similar classes.

To sum up: though Whitehead's account of the Method of Extensive Abstraction is not always stated very consistently, the inconsistencies are not as great as Grünbaum makes them out to be. Some of the difficulties can no doubt be traced to Whitehead's mode of exposition. Despite the disagreement among commentators, the interpretation of the method which seems to agree best with Whitehead's general position is that he was not trying to construct a geometry from sense-experience, but rather using a mathematical model to make clear certain relations appearing in perception. Such an interpretation would bring into harmony his earliest work on the theory of interpoints, his account of the method in P N K and C N, and also the P R account itself, which is a purely formal logical investigation.

There still remain the difficulties arising from the nature of the mathematical continuum—problems relating to the notion of super-denumerable infinities, etc., to which Grünbaum draws our attention. But as we have seen, the intuitionists are faced with similar difficulties, which they have endeavoured to meet. It may be, as Whitehead has already pointed out, that the assumption that physical space has the character of the real number continuum is merely one of convenience. Further, like the intuitionist, this conception may need reinterpretation on Whitehead's view. However, this is something Whitehead does not seem to have discussed, and one can only speculate as to how he would have dealt with this problem. From some of his assertions in M T, in which he takes up an operational view of mathematics, he would seem to be moving towards the intuitionist position.

[1] A E, p. 177.

SUMMARY

Some account is given of the Method of Extensive Abstraction and the criticisms which have been raised against it. Grünbaum, for example, argues that it leads to a number of difficulties. Since in modern geometry an interval is composed of a super-denumerable infinite number of points, Whitehead has to discover in sensed nature a super-denumerable infinity of abstractive sets to define them. Further, as there is no means of distinguishing perceptually the abstractive classes defining two distinct neighbouring points, ambiguity of convergence occurs.

According to Nicod, Whitehead's method could be taken not as an analysis of the real world, but as the construction of a pure geometry of volumes, which is translatable into a geometry of points, and for which, unlike points, a specific interpretation exists in sense-perception. As against Grünbaum's criticism of ambiguity of convergence, Whitehead could argue that in the intuitionist's continuum, too, one would not always be able to differentiate between two neighbouring points. Some of Grünbaum's criticisms arise from his own assumption that space and time have the continuity of the real number continuum.

The ancestor of the 'method' is the theory of interpoints, where a point is defined in terms of a class of entities (objective reals) having a similarity of position.

PART TWO

THE DYNAMICS OF WHITEHEAD'S PHILOSOPHY

ABSTRACT

IN this part we consider Whitehead's theory of feelings. By a 'feeling' he has in mind the particular process whereby the fundamental stuff (or events) of the universe is organized. The events in nature are pictured as exemplifying forms of energy having a rudimentary emotive tone. Whitehead conceives an event as being made up of (i) a focal centre which is a combination of the aspects of other events, and (ii) a field of influence—the modifications it sets up throughout nature. His account of simple physical feelings concerns itself largely with the transmission of forms of energy along physical routes of events. The character of each event in the route conforms to that of its predecessors. It is this which gives physical objects their character of endurance.

Whitehead has to explain why on his view we perceive not a series of events having electronic or molecular characteristics, but perceptual objects, each with its own particular kind of existence. He assumes that in sense-perception the change is brought about by a process of simplification. As a result of this process the characters illustrating the many individual events are fused into the one perceptual object. This leads us to discuss intellectual feelings, and to examine different kinds of perception, illusory and authentic, as well as belief and judgment. We next consider Whitehead's theory of societies, his view that the universe is built up of a vast changing society of events containing various subordinate societies which form the material bodies around us, and which themselves contain groupings of smaller societies. We also deal with the analogy drawn by Whitehead between direct experience and the notions of physics, and finally with his attempt to reconcile free will and determinism.

CHAPTER VIII

THE THEORY OF PREHENSIONS

I

THE theory of prehensions or feelings begins with a study of the simple causal elements in physical nature, which are termed 'simple physical feelings'. However, in Whitehead's system the concept of feeling, of which there are several varieties, also covers the reactions of biological organisms to their environment, as well as perception, cognition and judgment in man.

Before proceeding to deal with Whitehead's account of 'simple physical feelings' which, he claims, lays the foundation of the treatment of cosmology in the philosophy of organism, we need to indicate that there are two main types of physical feelings, (*a*) simple, and (*b*) complex. In the latter we deal with the elements in our perceptual field, sensory patterns which not only have a spatial extension, but also a temporal duration, since any observed event always takes a period of time to manifest itself.

In the case of 'simple physical feelings', we are rather concerned with the transmission of a form of energy from event to event in the physical world. This may seem very puzzling, but one should remember that Whitehead in his philosophy not only gives physical events a concrete existence, but he also believes they have certain affective characters which are similar in type to the affective elements we are aware of in experience; that they are in effect throbs of emotion. However, he is careful to point out that experienced emotion is not what he calls bare emotion, since it has already undergone a considerable amount of complex organization. 'But even so,' he tells us, 'the emotional appetitive elements in our conscious experience are those which most closely resemble the basic elements of all physical experience.'[1]

Whitehead seems to identify these throbs of primitive emotion which, he thinks, make up the events in the physical world, with the physical notion of quanta of energy. He holds the view that physical

[1] P R, p. 228.

science only deals with the spatio-temporal and quantitative characteristics of such events, that the concrete physical world has a certain qualitative affective content which the scientific account abstracts from. So when we sometimes find him speaking of 'the transference of throbs of emotional energy, clothed in the specific forms provided by sensa',[1] he is, strange as it may seem, simply referring to the transference of quanta of energy from event to event in the physical world.

Take, for example, the following passage. 'Thus there is a vector transmission of emotional feeling of a sensum from *A* to *B*. In this way *B* feels the sensum as derived from *A* and feels it with an emotional form also derived from *A*.'[2] Now though it might seem that Whitehead is describing here the process of conscious sense-perception, it is evident from the context that he is really dealing with the way physical events causally influence each other—in other words, with the transmission of a form of energy from *A* to *B*. When Whitehead describes such a physical causal process, he appears to equate the notion of 'energy' with 'emotional intensity', and the particular form this energy assumes, for example, an electro-magnetic wave, with a sensum.

As a result, he stretches the meaning of the words 'sensa' and 'emotional intensity' to cover entities and processes which seem to bear little resemblance to what we normally mean by these terms in sense-perception. Whitehead would undoubtedly defend his position by pointing out that these physical entities and processes are merely the primitive prototypes of the affectively toned sensory patterns directly observed by us in conscious perception. Hence, unless the reader knows the precise field (whether physical or perceptual) Whitehead has in mind when he uses these notions, confusion will inevitably result. And it is the failure to indicate their precise context of use which has wrecked most interpretations of P R.

It is clear that when we come to consider the notion of 'feeling', we have to put aside its usual associations. In our ordinary usage it generally refers to a specific type of experience going on within the mind or body of some human being. But we need to take great care not to interpret Whitehead on these lines, as the term functions quite differently within the context of P R. For example, he uses the phrase 'simple physical feeling' in a purely objective sense to refer to the way physical events causally influence each other.

[1] P R, p. 163. [2] P R, p. 446.

As we have seen, the model on which he bases his account of 'simple physical feelings' appears to be the notion of the transference of energy from event to event in physics, especially in electrical systems with radiation (where e.g. the electron is conceived of as propagating from itself a stream of influence). Whitehead draws an analogy between the transference of 'sensory affective character' from one event to another in sense experience (i.e. our awareness of a stream of sensory experience) and the transference of energy in physical systems.

He would claim that this correspondence arises from the fact that these physical notions are derivative abstractions from the concrete data given in direct experience. Such a correspondence is, however, rather due to the manner in which Whitehead describes experience than to the nature of experience itself, though it is not denied that there may be some resemblance between, for example, the experienced character of passage, and the flow of energy in electro-magnetic theory. Nevertheless in the latter case we deal with something largely hypothetical, a physical construct; in the former, with a directly observed character of our experience.

Now the above analogy gives us the key to the empirical side of the philosophy of organism. Whitehead has stated that he is looking for elements in experience in terms of which a systematic metaphysical cosmology can be constructed-elements not only found in direct experience, but also in physical nature. According to him, 'The key notion, from which such a construction should start, is that the energetic activity considered in physics is the emotional intensity entertained in life'.[1]

In this passage we have, as it were, the Open Sesame to his theory of feelings in P R. Not only does this theory apply to the basic elements within the physical world (simple physical feelings); it also applies to perceptual experience (complex physical feelings). It means that, in the case of sense-perception, Whitehead is really using the concept of the electro-magnetic field as a device for representing certain characters of the perceptual field. This no doubt explains why there is a resemblance between his view of perceptual experience and that of the Gestalt psychologists. They both replace a description in terms of static sense-data by a description in terms of dynamic process.

When Whitehead then speaks of 'simple physical feelings' in P R, he is not talking about the elements given in conscious sense-

[1] N L, p. 96.

perception, but of events in physical nature. He is referring to the transmission of a form of energy from one physical event to another. Whitehead therefore conceives a 'simple physical feeling' as a pattern of temporal transition, the transmission of a physical pattern from the present into the future.

This account of a 'simple physical feeling' is not, as might be thought, entirely new; we find something very like it in P of R. Whitehead there, when speaking of the way a mass-particle (e.g. an electron) transmits its field of influence into the future, points out that 'It is to be a limited atomic field with a foot in two camps, for it represents the property of the future as embodied in the past'.[1] And he adds 'the mass-particle is a transmission of physical character along its lines of force with a definite finite velocity'.[2]

It will be seen that the physical field streaming into the future is identical with Whitehead's conception of a 'simple physical feeling', which also has a foot in two camps, since it originates from a transmission in the past, and is itself a transmission of physical character into the future. It is worth noting here that Whitehead in M C took the lines of force of the modern physicist to be the 'ultimate unanalysable entities which compose the material universe', just as in P R 'simple physical feelings' are 'the ultimate elements from which a more complete discussion of the physical world—that is to say, of nature—must be derived'.[3]

Whitehead further tells us 'A simple physical feeling is an act of causation'.[4] The reason for this is that a 'simple physical feeling' resolves itself into such a stream of physical influence. We may regard the past event as the 'cause', for it transmits its character into the present, and the stream of influence from the present into the future as the 'effect', since it is conditioned by the past transmission.

All this seems fairly straightforward. 'But,' we are also informed, 'it is equally true to say that a simple physical feeling is the most primitive type of an act of perception, devoid of consciousness'.[5] This remark may seem a little puzzling. However, a hint is thrown out in S M W[6], where Whitehead notes that Bacon distinguishes between perception in physical nature and sense-experience in human beings. In the former case, Bacon mentions the action of loadstone on iron at a distance, and the ability of a weather glass to register the least difference of the weather in heat

[1] P of R, p. 75. [2] P of R, p. 76. [3] P R, p. 338.
[4] P R, p. 334. [5] P R, p. 334. [6] S M W, cf. p. 52.

or cold. The point made here is that the sensitivity of physical things to changes in their environment is in some ways analogous to sense-perception in human beings. Similarly, Whitehead uses the phrase 'physical perception' to refer to the process whereby one event causally influences another. Hence what he terms a positive prehension is in its simplest form really such a transmission of influence from event to event in physical nature.

So far we have only been considering an artificially simplified case of 'prehension', the linear transmission from one event to another in a historical route (making up, for example, some enduring physical object). We have therefore neglected the fact that each event is causally related to the rest of the events of the universe. According to Whitehead the physical character of any event is modified (to however slight a degree) by any other event, however separated by intervening events.[1] The relations of this route (or object) to the other events in its environment, which in a sense it excludes, as well as to the realm of alternative possibilities, are termed by him negative prehensions. On the other hand, the concept of positive prehension is used by him to refer to the inheritance of a definite physical character throughout a historical route (or enduring object).

Whitehead now makes the rather cryptic remark, 'The subjective forms of feelings depend in part on the negative prehensions'.[2] However, he is merely pointing out that the peculiar character of the present event (e.g. some definite electron) partly depends on the other events which form its environment, and whose characters may not seem to have any direct effect on it. Whitehead also believes that the alternative characters which that event could have exhibited still have some relevance for it.

What should be noted is that Whitehead here uses the term 'subjective form' to refer to a physical character (or scientific object) situated in some definite physical event. Hence he is not, as might seem at first sight, describing something given in introspective experience. By a 'subject' in this context, he usually means an event causally influenced by some other event in its past. Its particular patterned character is therefore conceived of as its 'form'. In the case of sense-perception, however, Whitehead uses the notion of 'subjective form' to refer to the complex patterns we are aware of in direct experience. This is another example of how careful we have to be to ascertain the precise context in which

[1] P N K, cf. p. 98. [2] P R, p. 335.

Whitehead's terms function, for they can apply either to the elements in physical nature or to human experience.

What Whitehead is discussing above is essentially the question of the endurance of physical objects. In an earlier work he defined physical endurance as 'the process of continuously inheriting a certain identity of character transmitted throughout a historical route of events';[1] or, as he puts it rather differently in PR, 'the subjective forms of the immediate past are continuous with those of the present'.[2] In other words, the character of each event in the route conforms to its predecessors, so that its specific character is inherited throughout; and it is this which makes us say we are observing the same object throughout a period of time.

Hence, according to Whitehead, a 'simple physical feeling' embodies (*a*) 'the reproductive character of nature', and (*b*) 'the objective immortality of the past'.[3] It embodies (*a*), since it is the transmission of character into the future which gives rise to the novel event; and (*b*), since it originates from the immediate past, from which this character has been inherited. The repetition of such an identical pattern in a route of successive events enables us to talk about physical objects having a continued existence.

II

The first three categorial obligations of Whitehead's theory of feelings may be regarded as the principles regulating the transmission of character from event to event in nature. The first category, he tells us, has to do with the self-realization of an event, by which he would seem to mean, despite his Spinozistic way of putting it, that an event exhibits the characteristic of change or passage. When Whitehead then says, 'An actual entity is at once the subject of self-realization, and the superject which is self-realized',[4] he refers to the fact that an event is a process of becoming in the present, and is also that which is about to become in the future. Or, putting it in the terminology of electro-dynamics which he is so fond of using, it is a present focus with a field of force streaming into the future.

He defines Category I as follows: 'The many feelings which belong to an incomplete phase in the process of an actual entity, though unintegrated by reason of the incompleteness of the phase,

[1] SMW, p. 136. [2] AI, p. 235. [3] PR, cf. p. 336. [4] PR, p. 314.

are compatible for synthesis by reason of the unity of their subject'.[1]

In simpler language, we may say that the many characters transmitted from the past, though as yet unintegrated into a definite event (or superposed in some region of space-time) are yet capable of being thus integrated, by reason of the unity of the perspective standpoint in the immediate future. Thus the unity of the future event (which he calls its final aim or superject) is already present as a condition (in the form of such a perspective) determining the transmission of character into the future. It hence determines the form of the novel event, since with the creative advance of nature the abstract region becomes a definite event.

Whitehead has to explain why, though each event is complex in character, i.e. made up of the aspects of the events in its past, it is yet a unity and not a mere aggregate of elements. This he supposes comes about as a result of the many characters transmitted from the past being combined or superposed within a region of space-time in the immediate future. The process of integration then seems to come to nothing more elaborate than the superposition of these transmitted aspects, and the resultant elimination of incompatible characters so that we arrive at the one pattern characterizing the novel event.

Whitehead states Category II, 'Objective Identity', as follows: 'There can be no duplication of any element in the objective datum of the satisfaction of an actual entity'.[2] By this he means that each element has an essential self-identity in the patterned event. The reason for this category becomes apparent if one remembers that on Whitehead's view each event, as it were, mirrors the universe, or unifies within itself aspects of other events into the one perspective characterizing it. Without some such condition, instead of a definite perspective, we would have a confused medley of conflicting aspects. Only those aspects having an identity of character are unified into the one perspective.

Category III, that of 'Objective Diversity', states that the diverse elements making up the pattern (sensory or physical) cannot coalesce together, since each character has to have an individual self-identity in that pattern. So this patch of red is necessarily different from and excludes the patch of green in some other part of the visual field (or in physics one electron is different from another).

Whitehead also emphasizes that each sensory (or physical)

[1] P R, p. 315. [2] P R, p. 318.

character will differ from other similar characters in accordance with the pattern in which it finds itself. He is putting forward 'the doctrine of real unities being more than a collective disjunction of component elements'. This view resembles the Gestalt doctrine that the whole 'set up' of the perceptual pattern affects or modifies its component parts. The component entities will thus differ in their intrinsic character according to the patterned whole in which they are ingredient. In physics this takes on the form of 'each molecule is affected by the aspect of this pattern as mirrored in it, so as to be otherwise than what it would have been if placed elsewhere'.[1] On Whitehead's view then, every sensory quality or scientific object has a character which differs in accordance with the pattern of which it forms a part. This doctrine fits in much better with our everyday perception that colours, for example, look lighter or darker when observed against different backgrounds.

When Whitehead further stresses that each element differs from similar elements occurring in other patterns, one is strongly reminded of Leibniz's 'identity of indiscernibles', since on both views no two elements can have exactly the same qualities. We have already seen that according to Whitehead the red perceived 'here and now' is not the same as that perceived 'there and then', as it is modified by the context in which it finds itself.

To sum up, these three categories impose conditions on the process of integration by means of which the transmitted aspects of past events are integrated into the novel event. Category I refers to the unity given to the transmitted characters from the past by the extensive standpoint in the immediate future to which they are transmitted. II refers to the self-identity of the elements within the unified perspective, due to the elimination of incompatible aspects. And III refers to the diversity of these elements within the pattern, a factor which gives rise to a contrast or patterned effect. These categories seem to be rather *ad hoc* logical postulates than descriptive of any natural laws.

III

In his discussion of the Category of Transmutation[2] (Category VI) Whitehead tells us that our usual way of consciously prehending

[1] S M W, p. 185.

[2] P R, cf. pp. 355-60. (Categories IV and V, of the set of categorial obligations, are discussed in the next chapter.)

the world is by transmuted feelings. This is another way of saying that the way we are consciously aware of the world is by means of perceptual objects. As we shall see in more detail in Chapter XI, in human perception, according to Whitehead, the physical characteristics belonging to the specific events making up the historical routes of events or objects in the external world are transmitted to the final percipient, by means of his sense-organs, in the form of discrete entities or nervous excitations. Whitehead thus has to postulate some category to account for the substitution of the one perceptual object or sensory quality in place of these component physical entities. He therefore assumes that the perceived sensory quality is a modified version of the multiplicity of physical forms of energy transmitted to the percipient. In a transmuted feeling, he tells us, a quality common to all the events of a route is transferred from the individual events to qualify the one region (or object) in the form of a predicate. As he puts it, in our perception of the route constituting, for example, a stone, 'the immediate percept assumes the character of the quiet undifferentiated endurance of the material stone, perceived by means of its quality of colour'.[1]

Whitehead claims that this process of transmutation or integration of the many transmitted physical characters of a route into one qualitative character, illustrating a spatial region, is not merely found in sense-perception. It is also found in physical nature, where the many microscopic electrons are averaged out to give rise to macroscopic ponderable matter.

On the level of physics, for example, the qualitative character would be the unit scalar charge derivative from the vector transmission of energy. We can therefore understand what Whitehead means when he describes the perceptual process in the following terms. 'In the language of physical science, the "scalar" form overwhelms the original "vector" form: the origins become subordinate to the individual experience.'[2] In other words, he assumes that in perception the simply located sense-qualities are derived from the external events which transmit their peculiar characteristics *via* the body to the final percipient. Whitehead further compares the clear-cut qualitative character in sense-perception with the 'qualitative intensity of localized energy' which 'bears in itself the vector marks of its origin', since he thinks that the perceived quality exhibits (in the shape of an affective tone) its

[1] P R, p. 107. [2] P R, p. 301.

manner of derivation from the physical and physiological transmission from the past.

We have already seen that, when Whitehead deals with the entities investigated by physical science, the notion of 'physical feelings' corresponds to physical vectors, and that of 'conceptual feelings' to the scalar characters or unit charges.[1] It is therefore not so odd for Whitehead to talk of inorganic bodies having 'conceptual feelings', or of transmuted feelings being unconscious. Only when his discussion centres on what he calls intellectual mentality can he be said to be dealing with conscious perceptions and judgments. Or, to take another example, when Whitehead speaks of physical events (i.e. those events which form the subject matter of physics) as having 'mental poles', he draws our attention to certain permanent characteristics which can be traced within the vector transmission.

Whitehead believes that the approach to conscious perception consists in the gain of a power of abstraction, so that the irrelevant multiplicity of physical detail is eliminated, and emphasis is laid upon the systematic characters pervading events. In this way we become aware of the world of common-sense objects illustrated by sights and sounds.

IV

In S M W Whitehead stated his theory of prehensions in more general terms. To avoid misunderstanding he made it abundantly clear that a 'prehension' has a complex character, that he thought of it as a process of integration (or unification) of the diverse aspects of other events into some particular pattern grasped into the unity of a perspective standpoint here and now. Hence, as we have already seen, for Whitehead an event is made up of (i) the pattern of aspects of other events integrated in its unity 'here' and 'now', and (ii) the pattern of modifications it sets up in the neighbouring events.[2]

This doctrine led Russell to assume that Whitehead was elaborating a mystic pantheism.[3] He is, however, saying little more than that an event is, after the fashion of a field of force, made up of (i) a focal centre, which is a combination of the aspects of other events (i.e. the event as in its own prehension), and (ii) its

[1] For a simple discussion of fields of force, cf. A E, pp. 220-8.

[2] S M W, cf. p. 129.

[3] *Analysis of Matter*, pp. 340-1.

field of influence, the modifications it sets up throughout nature (i.e. the event as in the prehension of other events). What is curious about Russell's criticism is that he himself at one time put forward a very similar view.

Whitehead makes himself more explicit when he goes on to explain what this notion of the interaction of events (or reciprocity of aspects) comes to. 'The ordinary scientific ideas of transmission and continuity are, relatively speaking, details concerning the empirically observed characters of these patterns throughout space and time.'[1] In other words, Whitehead conceives each event's prehension of every other event's pattern of aspects (the fact that they mutually modify each other) after the fashion of a transmission of influence from event to event throughout nature.

The chief reason given by both Russell and Whitehead for the adoption of their respective views is that they have been influenced by the field theory of physics. Using this as a model, Russell has constructed a theory of perspectives and Whitehead a theory of prehensions. As a consequence both also accept the physical notions of continuity and transmission through a medium. For there would be no sense in talking about a point of view, unless it stood in some form of relationship to the perceived object.

In Russell's account the relationship is a spatial one, and the particular character or aspect the spatial perspective assumes is due to the transmission of light from the object to this point of view. On Whitehead's theory, the mutual prehensions of events take the place of spatial relations, whilst he substitutes for the transmission of light the notion of 'physical prehension', i.e. the transmission of a form of energy (or sensory character) which he sometimes expresses in terms of perspectives. Instead of having a perspective space (or system of points of view), we have an interrelated structure of events (or manifold of prehensions). Each event or spatio-temporal point of view can be thought of as mirroring the universe, since it stands in perspective relationship to every other event.

Russell further told us in his account that two places in perspective space are associated with each aspect of a thing: the place where the thing is, and the place which is the perspective of which the aspect in question forms part. 'We may distinguish the two places as that at which, and that from which, the aspect appears. The "place at which" is the place of the thing to which

[1] S M W, p. 130.

the aspect belongs; the "place from which" is the place of the perspective to which the aspect belongs.'[1]

Though the above is expressed in terms of simple location, in terms of places, instead of events, it resembles Whitehead's view on a number of points. The 'place of the thing' to which the aspect belongs (i.e. at which it appears) resembles the aspect of some other event which the present event includes in its perspective standpoint. The 'place of the perspective' to which the aspect belongs (i.e. from which it appears) corresponds to the event or perspective standpoint in which the aspect is an element. Or, to take an example given by Whitehead from the field of perception, what we perceive 'is the perspective of the castle over there from the standpoint . . . here',[2] where 'here' refers to our bodily event.

In Whitehead's P R, we find a similar doctrine.[3] The two places associated with a given aspect correspond to the events concerned in a simple physical feeling, the 'place of the perspective' to (*a*) the subject of that feeling (or perspective standpoint), and the 'place of the thing' to (*b*) the initial datum (or the event in the past). What Whitehead calls the objective datum resembles the 'aspect', as it is the perspective of (*b*) from the standpoint of (*a*), or the way (*b*) transmits its character to (*a*). What must be remembered is that by a 'simple physical feeling' Whitehead is really talking about the transmission of a form of energy from event to event in physical nature rather than about conscious perception, and that he is trying to draw an analogy between it and our assumed perception of perspectives.

According to Whitehead, then, every event (or object conceived as a route of events) is to be thought of after the fashion of a field of force, having a focal centre from which emanates a stream of influence. The radiating field is the correlated system of aspects of that event (or object) as it modifies the other events in nature.

[1] *Our Knowledge of the External World*, 1949 ed., p. 100.

[2] S M W, p. 87.

[3] Whitehead analyses this process into five factors, (1) the 'subject', namely, the present event in which the perspective is an element; (2) the 'initial data', the events in the past, from which this perspective has been derived; (3) the 'elimination' in virtue of negative prehensions, the elimination of the incompatible aspects of these events by their exclusion from the prehended pattern; (4) the 'objective datum', the one perspective or pattern derived from the aspects of these past events, as the result of a process of integration (or concrescence); and (5) the 'subjective form', the integrated perspective as characterizing the present event. (Cf. P R, p. 312.)

Each event or focal centre is therefore a composition of the overlapping aspects (or fields) of all the other events in the universe, whilst its aspects in other events are the set of modifications set up by it. Thus he states, 'we must admit that within any region of space-time the innumerable multitude of these physical things are in a sense superposed. Thus the physical fact at each region of space-time is a composition of what the physical entities throughout the Universe mean for that region.'[1]

The integration (concrescence) of these aspects is made possible by their being thus superposed in a certain region of space-time, so that incompatible aspects are eliminated, just as incompatible forces cancel each other out. Whitehead himself draws such a comparison when he states: 'The "datum" in metaphysics is the basis of the vector theory in physics; the quantitative satisfaction in metaphysics is the basis of the scalar localization of energy in physics.'[2] That is to say, he compares these transmitted aspects of other events to vectors, as, though they are in the standpoint here, they have an essential reference to other places and times; whilst he compares the derivative patterned event with the scalar localization of energy.

However, an event is not merely a bare region of space-time; we are also aware, for example, in perception that we observe aspects of things (or sensory patterns) around us, i.e. trees, houses, etc. When Whitehead then goes on to speak of conceptual feelings, one of the things he has in mind is just these sensory aspects. He therefore distinguishes between physical and conceptual prehensions; in this case between the spatio-temporal aspects of other events, integrated into the present event, and the aspects of their specific sensory qualities which characterize this event. Since each event is made up of these two kinds of prehension, he terms it a 'hybrid' prehension. This would seem to resemble his earlier analysis of nature in terms of events and sense-objects. Further, according to Whitehead, in physics (or the physical world) the notion of a physical prehension takes on the form of the transmission of energy from event to event, whilst 'hybrid' prehensions refer to the origination and direction of energy, i.e. the way one type of energy is changed into another.

[1] A I, p. 202. [2] P R, p. 163.

V

From the foregoing one can clearly see the resemblance between Whitehead's view and that of Russell, though Russell thought that Whitehead's view was incompatible with science. Both accept the notion of the continuity of physical processes, as seen, for example, in physical field theory and the transmission theory of light, where such a continuous linking up of happenings is assumed, in one case of electro-magnetic effects, in the other of optical effects. This notion seems implicit in Russell's view when he illustrates his theory of perspectives by a photographic example. He tells us: if we assume, as science normally does, the continuity of physical processes, we are forced to conclude that, at the place where the plate is, and at all places between it and the star which it photographs, something is happening. This is due to a process which radiates outward from the star, so that in every place at all times a vast multitude of things must be happening. The happenings or aspects scattered throughout space-time are then apparently due to the transmission of light from the star to these places.[1]

The whole theory of perspectives put forward by Russell thus, as he admits, presupposes the transmission of light through a medium. It assumes, at least, that there is a direct relationship between the object and what are called its aspects (otherwise there is no sense in terming the latter a perspective). It also assumes that this relationship can be extrapolated throughout space and time.

However, 'the principle of the continuity of physical processes', involving as it does a crude metaphysical doctrine of internal relations, with its extrapolation of effects throughout space-time, remains even in physics a questionable hypothesis. This principle, as used in Russell's theory, seems little else than a convenient device or rule for putting some kind of order into our apparently divergent perceptions of things, which are not really as divergent as Russell would have us believe. In its essentials (if we neglect the causal implications), it assumes that these perceptions stand in projective relationships to each other, and also to the things perceived; that every possible observation made at every point of space will vary in such a geometrical fashion. In perception the limitation of this view becomes apparent, since, as the Gestalt psychologists have shown, we do not perceive objects in the way

[1] *Analysis of Mind*, cf. pp. 99-100.

the geometrical laws of perspective would lead us to expect. There is a much greater constancy about them than some philosophical writers have led us to believe.

Russell considers a piece of matter as a first approximation to be 'the collection of all those correlated particulars which would normally be regarded as its appearances or effects in different places'.[1] But by 'appearances', he pointed out, he did not mean anything that must 'appear' to somebody, 'but only that happening, whatever it may be, which is connected, at the place in question, with a given physical object—according to the old orthodox theory, it would be a transverse vibration in the aether'.[2] In other words, it need not be, as has been supposed, similar in type to our sense-perceptions, i.e. an unsensed sense-datum, but may also be an electro-magnetic effect; so that when an observer is placed at that point of view he perceives e.g. a particular colour. Thus the set of particulars making up the object will in some parts of space merely include physical effects, whilst in other parts, where there are brains and sense-organs, it will include physiological and psychological effects.

Russell apparently justified such a theory of perspectives by saying that he based it on the notions of modern physics. 'To my mind, the world is full of particulars of the sort dealt with by physics, and some of these particulars (namely, those in places where, as we say, there is a brain) have peculiar effects which are called "being known" or "being experienced".'[3] In other words, he supposed that when, for example, we say we see a penny, what we really perceive is a member of a certain system of particulars, and it is that member which is situated in a specific part of our brain. There is a continuity of effects of particulars from the penny (or set of effects) in the physical world, to our sense-organs, brains and psychological experience. Thus the penny is really such a system of correlated physical, physiological and psychological effects.

He therefore believes that this theory harmonizes physics, the physiology of the sense-organs, and psychology. And he adds that 'My view of the relation of what we perceive to physics is the same as that of Dr Whitehead, who first persuaded me to adopt it'.[4] As we have noted, Whitehead's position in SMW (insofar as he deals with the causal account) closely resembles this view, since he

[1] *Analysis of Mind*, p. 101.
[2] Ibid, p. 101.
[3] *Mind*, 1922, p. 482.
[4] Ibid, 1922, p. 483.

seems to hold that there is a similar continuity between the physical world and the perceived field.[1]

SUMMARY

By the phrase 'simple physical feelings' Whitehead refers to the transmission of a form of energy from event to event in physical nature. He regards these entities and processes in the physical world as the primitive prototypes of the affectively toned sensory patterns directly observed by us.

Simple physical feelings, we are told, embody (*a*) 'the reproductive character of nature'—the transmission of character into the future which gives rise to novel events, and (*b*) 'objective immortality', since these events originate from the immediate past and live on, as it were, in the present. The repetition of such an identical pattern in successive events enables us to talk about physical objects as having a continued existence. In order to explain how the multiplicity of transmitted aspects of other events is integrated into a definite space-time perspective, Whitehead brings in Categories I, II, and III, whilst the Category of Transmutation is introduced to account for the substitution of one perceptual object (or physical object) in place of the many routes of physical events occurring in nature.

Russell's theory of perspectives and Whitehead's theory of prehensions are compared. Both seem to use the physical field as a model for their respective theories.

[1] That there is a remarkable family resemblance between their respective views has been noted by McGilvary, pp. 234-6 of the Schilpp volume. It is also interesting to note that Russell in his answer to Strong (*Mind*, 1922, p. 480) repudiates the view that he is a phenomenalist. 'In fact, however, I am not a phenomenalist. For practical purposes, I accept the truth of physics, and depart from phenomenalism so far as may be necessary for upholding the truth of physics.'

CHAPTER IX

PERCEPTION AND PROPOSITIONS

I

WHITEHEAD repeatedly criticizes philosophical discussions of perception, where, he claims, the Aristotelian substance-quality doctrine still holds sway. As a doctrine, Whitehead thinks it owed its genesis to the Greeks' excessive preoccupation with the most sophisticated form of perception, namely that given in our structured perceptual field: the kind of perception he calls 'presentational immediacy'. But Whitehead, accepting as he does a causal theory, assumes it to be the end-point of a long chain of physical and physiological occurrences.

When we observe the sharp-cut many-hued world given in perception, we tend to overlook the actual functioning, the activity which gave rise to it. And yet Whitehead claims we discover it in experience, as a feeling of vague bodily awareness which he calls perception in the mode of 'causal efficacy'. It is through this mode of perception that we obtain information of the way in which our sense-perceptions have been derived from the physical world (i.e. through bodily functioning). As he says of our perception of a blue patch, though it appears as the most simple component in experience: 'Nothing can be more simple or more abstract. And yet unless the physicist and physiologist are talking nonsense, there is a terrific tale of complex activity omitted in the abstraction.'[1] His view is that when we observe a specific sense-datum there is always in the background the body of the observer and the physical world beyond.

Whitehead, as we have seen, is of the opinion that during their passage through the body of the observer, the physical forms of energy become transformed into the clear-cut sensory qualities which mark out definite contemporary regions of space. Though in this connection Whitehead speaks of the sensa as being 'projected' on to these regions, he does not think they are, as it were, literally

[1] M T, p. 166.

shot out of our heads into external space. He considers the perceived spatial regions to be just as abstract as the qualities characterizing them. They are both simplified editions of the events in physical nature.

According to Whitehead, we arrive at these entities, the 'spatial region' and the 'qualitative sensory character', by emphasizing certain details of events and suppressing others. (Of course, it should be remembered that this process takes place on a purely unconscious level, as we are aware only of the end-product in conscious experience.) What is immediately given to us in 'sense-reception' (as opposed to sense-perception) is a physical event or route of events, marked out by physical forms of energy, which affects our bodily functioning. And as a result of an elaborate process of analysis and emphasis we arrive at our common-sense world of perfectly defined objects, situated in definite regions of space and illustrated by diverse sense-qualities.

If we start from Whitehead's premise that the forms of physical energy are really pulses of emotional experience, then there is of course no logical reason why sensory qualities should not be, as he claims, a more intense form of experience derived from them.

II

We come now to consider Whitehead's discussion of 'propositions' and 'propositional feelings' and their relationship to perception. A proposition in this particular context plays a somewhat different role than it usually does in most philosophical discussions. Whitehead conceives it as a combination of the structural elements we have been talking about: (*a*) the regional space-time position (the logical subject) and (*b*) the eternal object or sense-quality (the predicative pattern). The qualification of (*a*) by (*b*) gives rise to a perceptual proposition, namely some specific sensory quality located at a definite place and time.

Despite Whitehead's unusual use of the term 'proposition' to refer to perceptual data, this usage is not entirely unjustified. Whitehead considers the eternal object (predicative pattern), for example, 'blue', to be an attribute which we can express in the form of ϕx (or x is blue). If we substitute for the variable x some definite spatio-temporal region (or logical subject), thus giving the predicate a definite spatio-temporal determination, we then deal with a particular quality characterizing a specific place and time. One can thus readily understand why Whitehead calls such a characterized region a proposition. In a sense it would seem to be a translation of the Aristotelian S-P form of proposition on to a perceptual basis. Indeed Whitehead might claim that the logical concept is derivative from the perceptual one.

Further, according to Whitehead, not only do we think symbolically, but we also perceive by means of symbols. The coloured shapes we observe in our perceptual field are merely symbols for the primitive feeling elements in our experience which, he believes, give us direct awareness of the physical events in nature. For example, we may be walking along a street, when suddenly we observe in front of us a black patch surmounted by a smaller pink one. In this case we are aware of what Whitehead calls a proposition (or perceptual object) whose component elements are the sensory qualities black and pink, characterizing a specific spatial region. But our perception of such a coloured shape may set up in us a definite train of behaviour. We may go up to it and say, 'Fancy meeting you here, Jones', and indulge in a lengthy conversation about the vicissitudes of the English climate. Now Whitehead thinks it is frankly unbelievable that we should act in this manner if all we experience is a coloured shape. He believes that the reason why we behave in this way is due to our experience in the mode of 'causal efficacy' which gives us immediate acquaintance with the physical facts, against which we check the truth or falsity of our sense-perceptions.

We can now understand what Whitehead means when he says that the primary function of a proposition (or a theory) is as a lure for feeling,[1] though this may seem a pretty odd doctrine if one gives the word 'proposition' its generally accepted connotation. Apparently the function of a proposition is to refer to the emotional background of fact which gives it its significance. Our common-sense world as it appears in our perceptual field is to be taken as a symbol for the throbbing world of physical activities. The application of this view to statements about sense-perceptions can be seen from the following example. According to Whitehead, when we say we see a 'grey stone', the 'grey' refers to the grey shape immediately before our eyes. But the word 'stone' has reference to other elements in our experience, the feeling of the efficacy of the series of physical events which make up the stone

[1] P R, cf. pp. 260-1.

in nature. To use Russell's terminology, in perhaps the opposite sense to that in which he intended it to be used, it might be said that perceptions in the mode of 'presentational immediacy' (i.e. our awareness of elements in our perceptual field) give us knowledge by description of the actual physical world, whereas we get knowledge by acquaintance through our feelings in 'causal efficacy'.

To be aware of a proposition appears at first sight to be a rather complicated business for Whitehead, involving what he calls an 'affirmation-negation contrast'.[1] This requires some explanation. When Whitehead speaks of eternal objects or qualitative characters as considered by themselves in thought, he tends to describe them as negations, as we are merely dealing with abstract class concepts. Thus he tells us 'Conceptual feeling is the feeling of an unqualified negation; that is to say, it is the feeling of a definite eternal object with the definite extrusion of any particular realization'.[2] For example, we can only be aware of blue by observing blue things; when blue is considered by itself (in abstraction from fact) it becomes merely an abstract concept. In a proposition, however, the eternal object (attribute) is restricted to qualifying some specific region, and, since it still retains its abstract character, it might be described as a 'qualified negation', i.e. a predicate which has had its range of application limited to some definite spatio-temporal position.

'Propositions' taken as elements in our perceptual field (i.e. as percepts in the mode of presentational immediacy) can be said to form the negative side of the affirmation-negation contrast. On the other hand, the affirmative side can be said to be the efficacious background of feeling which he considers as an affirmation of the given fact, since it is the way in which we are directly aware of the physical world through 'causal efficacy'. Because of this, 'percepts' in the mode of 'presentational immediacy' are, to use his expression, 'theories' about actuality; they may apply to it, i.e. adequately symbolize facts, or they may not: they may be erroneous, thus giving rise to illusory perceptions. In an 'affirmation-negation' contrast, we then bring into existence a new product. To take the case of the 'grey stone'; our perception of a grey shape in our perceptual field and our awareness of the stone as an efficacious element are synthesized into one experience. We perceive not only a grey shape, but a grey stone of whose efficacy we have direct awareness. This doctrine is not altogether new. In his earlier work

[1] P R, cf. p. 377. [2] P R, p. 344.

Whitehead spoke of our awareness or apprehension of events as involving direct awareness of the passage and activity of nature. He also pointed out that knowledge of the particular characteristics of events depended upon our powers of comparison and is obtained through sense-recognition.

Despite the apparent phenomenalism of his earlier work, a comparison with P R shows that sense-awareness corresponds to 'causal efficacy', and sense-recognition to 'presentational immediacy'. In C N he made the further point that perception contains two elements, (a) thought, i.e. the abstraction involved in the recognition of sense-qualities, and (*b*) that which is not given in thought, i.e., nature as given in sense-awareness.[1] Indeed, in a footnote to P R[2] Whitehead refers us to the connection between his present position and his views on this subject in C N, where he tells us that in perception the logical subject is no longer a factor in fact but is reduced to a bare 'it'.

Whitehead's doctrine of perception as elaborated in his earlier writings is summed up by him as follows: 'Perception involves apprehension of the event and recognition of the factors of its character';[3] and this seems to correspond to the 'affirmation-negation contrast' (or conscious perception) of P R. As regards the derivation of the fundamental entities in experience and the manner in which they are related in perception, Whitehead's later view would then seem to be a direct development of his earlier one.

III

Let us consider in more detail the conditions determining the appearance of a perceptual proposition.[4] According to Whitehead, it depends upon (*a*) a physical feeling (i.e. bodily event or external event) from which are derived the necessary logical subjects or definite space-time regions, (*b*) a physical feeling which has among its components certain eternal objects or qualitative characters, and (*c*) an eternal object (or conceptual feeling) obtained from (*b*) by what he terms Conceptual Valuation (Categorial Obligation IV)[5]—the process whereby these characteristics are derived from the physical events.

It will be noted that Whitehead places the events (or physical feelings) from which the logical subjects and eternal objects are

[1] C N, pp. 3-4. [2] P R, p. 365. [3] C N, p. 189.
[4] P R, cf. pp. 368-70. [5] P R, pp. 350-2.

derived under two different headings, instead of only one as might be expected. This arises from the fact that, though in normal perception the logical subjects (spatio-temporal regions) and eternal objects (qualitative characteristics) are usually derived from the same physical event, they may, for example in hallucinations, be derived from two different sets of events. In such hallucinatory cases the qualitative characteristics are derived from centrally generated events, and the spatial regions of which they are predicated from the external events now happening.

However, even in normal perception some of the perceived qualities may not have been derived directly from the same event as the regions they characterize. This is covered by another condition (*d*), some conceptual feeling derived according to Categorial Obligation V (Conceptual Reversion),[1] involving another eternal object as its datum.[2] In other words (though one must remember that this account is also applicable at a physical level) our direct perception of certain sense-qualities may lead us to be aware of other (reverted) qualities appearing as a subconscious response to these sense-data.

Whitehead's earlier view[3] was that such reverted sensa (or images) play an important part in normal perception. Thus we may see a piece of velvet and imagine we are feeling it. Or we may say we see a dog, though all we really see is a certain patch of colour with indefinite boundaries. However, the dog we have in mind is more than that; it barks, its coat yields to our touch, it answers to a name, it bites if we tease it, and has other qualities of which at that moment we are not directly aware. As a general rule, a number of such qualities are 'instinctively' conveyed to us, thus reinforcing our perception of the coloured shape. We are therefore led to believe that there exists such a permanent collection of qualities (or substantial object) and that it is this perceptual object which we call the dog. Indeed in his earlier work he defined a perceptual object on these lines: 'The peculiar fact, posited by the sense-awareness of the concurrence of subconscious sense-objects, along with one or more dominating sense-objects in the same situation, is the sense-awareness of the perceptual object'.[4]

What Whitehead would seem to be doing here is making the old distinction between sensation and perception: that we reinforce the fragmentary sense-data by producing such 'subconscious

[1] P R, pp. 352-3. [2] P R, cf. p. 368.
[3] Cf. P N K, 24.1, 24.2. C N, pp. 154-5. [4] C N, p. 155.

sense-presentations', so that we perceive well-defined perceptual objects. But, as the Gestalt psychologists have shown, such a view depends upon the assumption that there is a point to point correspondence between the sensory stimulus and the actually perceived data. However, for naïve experience the subconsciously produced sense-qualities are just as much part of the immediately perceived field as are the directly given sense-qualities. It is only when we allow physiology to dictate to us the nature of what we see that we need to bring in a subconscious process to explain why we do not perceive what on such a theory we ought to. But, to do justice to Whitehead, he does emphasize that such qualities are on the same objective perceptual level as the immediately given sense-presentations, and that they only differ from them in their mode of causation.

In PNK Whitehead describes the processes involved in the appearance of a perceptual object as follows: 'Thus in the completed recognition of a perceptual object we discern (i) the primary recognition of one or more sense-objects in the same situation, (ii) the conveyance of sense-objects by the primary recognitions, (iii) the perceptual judgment as to the character of the perceptual object, which in turn influences the character of the sense-objects conveyed'.[1] In PR the completed recognition would seem to correspond to the 'satisfaction', (i) to Conceptual Valuation (Cat. IV), (ii) to Conceptual Reversion (Cat. V). As for (iii), perceptual judgments would seem to have at least two functions: (*a*) they deal with the compatibility of the 'recognized' and 'conveyed' sense-objects for the formation of some specific perceptual object, and (*b*) they give rise in us to the belief that the situation where we observe these qualities is an active condition for their recognition by us and other observers, i.e. is the situation of a physical object.

Thus we may walk into a room and see an arm-chair directly in front of us, and take it for granted that there is such an object facing us. We may even go up and try to sit on it, when to our surprise we find we have been looking into a mirror. And we have been deceived, Whitehead would say, not because the coloured shape located at the back of the mirror is different from that which we observe when we turn round, but because the conveyed sense-qualities have led us to believe that there was an object where in truth there was none. In such cases where the judgment under (*b*) is false, we say that the perceptual object is illusory.

[1] PNK, cf. Section 24.3, p. 89.

In P R, (iii, *a*) appears to be covered by the Category of Subjective Harmony (VII)[1] which deals with the way in which the subject gives rise to the 'conveyed' sense-data so that they are compatible with those derived from the external events and form a harmonious configuration. To take the case of the dog once more, the conveyed qualities, that it can bark, will bite if we tease it, etc., are such that they are compatible with the coloured patch we observe, and form together a complex qualitative pattern. Further, as we have seen, this qualitative pattern is integrated with the emotional background (i.e. feeling of physical events) to give us the contrast which is our conscious perception of the dog. The integration of these two perceptual elements would for Whitehead be a very simple form of intuitive judgment, and it would seem to resemble (iii, *b*). Such an intuitive judgment may of course be false, since the dog may only be a mirror image.

From the above examples we observe that the truth or falsity of perceptual propositions is relative to the subject who is aware of them, and is also dependent upon the manner of their derivation, whether directly from external events or by some mode of bodily functioning. For example, if we were now to see what appeared to be a tiger in front of us, we should have good grounds for believing it was a delusion. But on the other hand, if we were in some Indian jungle our reaction would be a very different one.

IV

Whitehead distinguishes four different kinds of propositional feelings, two main types and three derivative. We give the following table:

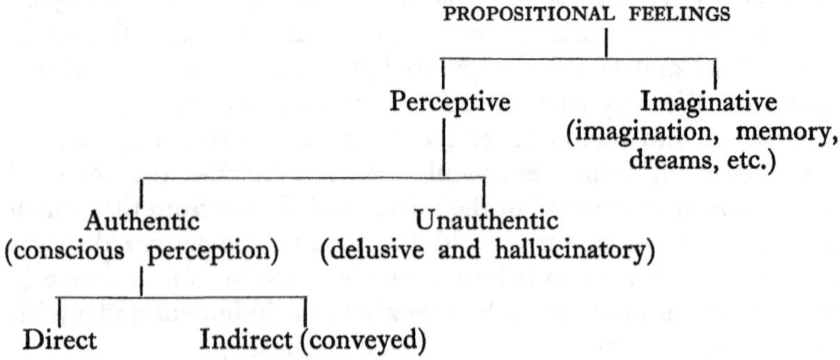

[1] P R, cf. pp. 360-1.

Propositional feelings are thus divided into two main types: (1) 'perceptive', (2) 'imaginative'. At first sight they would appear to resemble Hume's division into impressions and ideas. But, since Whitehead accepts a causal theory of perception, Humean impressions are for him already highly complex, and correspond to his 'direct authentic perceptive feelings'. This covers our awareness of a proposition, in which both the qualitative character, for example red, and the place and time at which red occurs, are derivative from the same event, and which, as related to the emotional background of physical fact, takes on the character of an 'impression of sensation'.

In the case of 'imaginative feelings', the qualitative characteristics are usually centrally generated, instead of being derivative from the external physical events. Unlike 'perceptive feelings', imaginative feelings involve two physical feelings: (1) that originated by the subject himself, from which the relevant qualitative characters are derived, and (2) the external physical events, which give us the particular place and time to which the qualitative pattern refers. Such characteristics, as we know only too well from our experience of day-dreaming, need not have and usually have not a very close relevance to the times and places about which they are predicated. Propositions, especially those entertained during our reading of imaginative literature, 'are tales that might be told about actual entities'.[1] We can now see what Whitehead means when he says that propositions are not primarily for purposes of judgment, that it is more important for them to be interesting than true; he is referring to their imaginative use when judgment is usually at a discount.

There may be degrees of difference between the characteristics of these two sets of physical feelings ranging from near-identity to cases where they differ widely.[2] Consider first the case where they may be almost identical, e.g. in memory, as when at night when dozing off to sleep we review the events of the day. It should be noted that, though the experienced qualitative patterns may greatly resemble, or for that matter be qualitatively identical with, those we perceived earlier, they are not directly due to the same events. In one case they are entertained by us now, in the other they occurred hours ago. But Whitehead would claim that when we reflect upon the day's events we are predicating certain centrally generated qualities of particular places and times, which

[1] P R, cf. pp. 370-2. [2] P R, cf. pp. 371-2.

in relation to ourselves we locate as being in the past. On the other hand, the characteristics of the two physical feelings involved in an 'imaginative' feeling may greatly differ from each other, as when e.g. we indulge in flights of fancy. The imagined characteristics may then have little in common with any present or past state of affairs.

As we have noted, we may be directly aware of a particular object in perception (i.e. through 'perceptive feelings') or imagine it at times when it is not directly present to our senses (i.e. through 'imaginative feelings'). To illustrate how the experience of a proposition can differ for different people, and even for the same man at different times, Whitehead takes as an example the proposition 'Caesar crossed the Rubicon'.[1] He first considers one of Caesar's soldiers who is prehending it in the form of a 'direct perceptive feeling', i.e. observing Caesar's crossing of the Rubicon. In this case, both the time and place and the predicated qualitative pattern have been derived through one physical feeling—his direct awareness of this specific event. Next Whitehead considers the same soldier who at a later date revisits the river and is reminded of Caesar's crossing. He now entertains the proposition in the form of an 'imaginative feeling', locating relatively to himself the past occurrence, and predicating of it an analogous qualitative pattern. Though this proposition is similar to the one he was directly aware of in the past, yet owing to its different mode of derivation his attitude towards it will be radically different. It is experienced with perhaps a strong feeling of belief, whereas the other is accepted as a direct fact of perception.

From what we have already said, it will be noted that the four different kinds of propositional feelings shade off into each other. Examining first the three types of perceptive feelings (two of which have yet to be dealt with), we see that 'direct perceptive feelings' shade off into 'indirect' ones as a result of the active conveyance of relevant sense-qualities. These in their turn blur into 'unauthentic' ones, as in this case sense-qualities are produced which have no immediate relevance to the situation they characterize. They give rise to delusive perceptions ranging from mirror images to wild hallucinations.

'Unauthentic perceptive feelings', as we have seen, tend to shade into 'imaginative feelings'. However, there is always this difference; in 'imaginative feelings', the imagination, over which the subject has some control, is within limits free to roam at will.

[1] P R, cf. pp. 276-7.

'Unauthentic' feelings are on the other hand due to what Whitehead calls a 'tied imagination',[1] as, with perhaps the exception of wild hallucinations, the conveyed sense-qualities usually have some relevance to the situation in which they appear. Further, the subject has little or no control over their appearance, since they usually arise from abnormal physiological conditions.

Let us examine more closely 'indirect perceptive feelings', our awareness of 'conveyed' sense-qualities, produced by the directly observed sense-data. For example, when we perceive, say, a coloured shape and a noise, we believe that they exist in the place where we actually see them. Now according to Whitehead we only take up such an attitude towards an 'indirect perceptive feeling' if the reverted sense-qualities have been 'transmuted', in other words, if they are taken as essential constituents of the perceptual object we are observing, when, as he puts it, they are felt, 'as if they had been physical facts in the nexus'.[2]

Thus to take the case of the dog again, we may hear a bark and see a coloured shape, and assume that there is an actual object, namely the dog, where we observe these sense-qualities. We therefore believe that the whole complex qualitative pattern, which we call the dog, is a characteristic of the actual events, whereas only the coloured shape and the bark have any right to this claim.

Now this transmutation of 'reverted' sense-qualities may lead us into error, for we may suppose that we are aware of objects when they are really not directly present to our senses, as in the case of mirror images. Nevertheless, our usual method of observing the surrounding world is through these 'indirect perceptive feelings'. We believe that such persistent qualitative patterns or material objects exist, even if we are not there to see them.

'Unauthentic' or 'delusive perceptions'[3] may be said to occur when the 'reverted' sense-qualities which appear in our perceptual field have little relevance to the characteristics of the external events, for example, when a dipsomaniac sees pink rats or elephants. However, the logical subjects (the spatio-temporal situations in which they appear) and some of the perceptual objects, the tables and chairs in the room, have usually some direct relevance to the actual happenings in the physical world. The complex qualitative pattern which characterizes the perceived spatio-temporal region may thus include sense-qualities derived through 'direct', 'indirect' and 'unauthentic' perceptive feelings. Whitehead would,

[1] P R, cf. p. 372. [2] P R, cf. p. 380. [3] P R, cf. pp. 381-2.

however, classify the awareness of this complex pattern under the heading of 'unauthentic perceptive feelings'.

It may be of help here to refer to Whitehead's treatment of 'normal' and 'delusive' perceptions in PNK and CN.[1] He there makes the distinction between 'active' and 'passive' conditioning events. Active conditioning events are causal for our perception of the sense-objects; usually they are part of a historical route (or physical object) in the physical world; the passive events are the rest of the related events of nature. In cases of normal perception, the events from which the spatio-temporal regions are derived, and which are characterized by the sense-objects, are generally the active conditioning events. The conveyed sense-objects are not directly obtained from the same events as the regions in which they are located, since the active conditioning events for their appearance are to be found in the observer. The events which they mark out are then to be regarded as the passive conditions.

To take a case of illusion, when an observer sees the image of a blue coat in a mirror and thinks it is a coat, Whitehead would say that what he sees is a blue coat and not a patch of colour. 'This', he concludes, 'shows that the active conditions for the conveyance of a group of subconscious objects by a dominating sense-object are to be found in the investigations of the medical psychologists.' And by this Whitehead means that when the percipient experienced the patch of colour he erroneously made the habitual response to the more normal occurrence—the direct perception of the blue coat. Consequently he took the event where he assumed the blue object was situated as a causal condition for the awareness of this object, whereas it was merely a passive one.[2]

[1] PNK, 23.3, 24.6. CN, pp. 151-155.

[2] Propositions are not something new in Whitehead's philosophy. In PNK he treats of them in the chapter on Figures. He is there concerned with the location of a sense-quality in a definite region; this he calls a sense-figure. He defines it as follows. 'The figure for a time system α, of sense-object O in situation σ is the relation holding, and only holding, between O and any α-volume congruent to a member of the set of α-volumes of σ.' (PNK, cf. p. 191.) That is to say, the abstractive class (or series) of these volumes converges to the ideal abstractive element, the timeless three-dimensional volume, which is the figure of the sense-object; and this can be said to be the logical subject of which this object is predicated.

SUMMARY

Whitehead thinks that the clear-cut sensory qualities marking out the contemporary regions of space in our perceptual field are simplified editions of the events in physical nature. A proposition on this account is to be thought of as having a perceptual rather than a logical significance. It is conceived as a combination of (*a*) the regional spatio-temporal position (the logical subject) and (*b*) the eternal object or sense-quality (predicative pattern), and seems to be a translation of the *S-P* form of proposition on to a perceptual basis.

Propositional feelings are divided into (1) perceptive, which may be either (*a*) direct, (*b*) indirect, or (*c*) unauthentic, and (2) imaginative. In (2) the qualitative patterns are usually centrally generated instead of being as in (1) largely derivative from the external events. The truth or falsity of perceptual propositions is relative to the subject who is aware of them, and to their manner of derivation. Perceptual propositions would seem to appear in P N K under the heading of sense-figures.

NOTE

Sense-objects

We need to enquire what Whitehead meant by a 'sense-object'. One thing is clear. He is not referring to what is usually called a sense-datum. The sense-object 'red', for example, is a purely logical entity which can be symbolized in the form '*x* is red', where *x* refers to the thing which may be red (or have this defining characteristic).

We might perhaps trace the origin of the above notion. In A E, for example, he made it quite clear that he considered a single self-identical sense-object to be already 'a phantasy of thought'.[1] Whitehead tells us that we arrive at the notion of a sense-object as a 'result of an active process of discrimination made in virtue of the principle of convergence. It is the result of the quest for simplicity of relations within the complete stream of sense-presentation.'[2] A sense-object is then not a perceived event manifesting a sensory character, but is related to it, in much the same way as on his view a mathematical point is related to a physical point.

[1] A E, p. 188. [2] A E, p. 192.

In C N he stated that by a 'sense-object' he meant a particular sort of colour such as Cambridge blue. But he was careful to point out that he was not talking 'of a particular patch of blue as seen during a particular second of time at a definite date. Such a patch is an event where Cambridge blue is situated.'[1] He is talking of the blue itself, which as a class-concept is expressible in the form '*x* is Cambridge blue', and he is not referring to any particular observed sensory experience. According to Whitehead, the particular sensory character that we are directly aware of in perception, becomes for thought 'merely the guardian of its individuality as a bare entity. Thus for thought "red" is merely a definite entity, though for awareness "red" has the content of its individuality. The transition from the "red" of awareness to the "red" of thought is accompanied by a definite loss of content, namely by the transition from the factor "red" to the entity "red".'[2] What Whitehead means by this is that a sense-object taken as a class-concept, though devoid of perceptual content, has a certain individuality about it; i.e. it has a certain defining characteristic, which refers to any event in which red appears. The relation between the entity 'red' (i.e. sense-object) and the factor 'red' (characterized event) is then that between a class to which many things may belong and one of these particular things.

In P R, where he classifies sense-objects under the heading of 'eternal objects of the subjective species', Whitehead asserts that any attempt to understand eternal objects in complete abstraction from the actual world results in reducing them to mere undifferentiated nonentities. He points out that we cannot know what red is by merely thinking of redness, but only by being aware of red things in our direct experience.[3] It is evident that eternal objects when taken by themselves (i.e. abstracted from matter of fact), have no actual 'being', since they are bare conceptual entities. As he tells us, an eternal object 'is referent to any actual entities, where the *any* is absolutely general and devoid of selection'.[4]

[1] C N, p. 149. [2] C N, p. 13. [3] P R, cf. pp. 362-3.
[4] P R, p. 364.

CHAPTER X

CONSCIOUSNESS, BELIEF AND JUDGMENT

I

FOR Whitehead, consciousness only appears in the more complex types of experience which he puts under the heading of 'intellectual feelings'. He distinguishes two types, (*a*) conscious perceptions, (*b*) intuitive judgments.[1] These might be said to be analogous to sensations and images. They are the names given to the 'perceptive' and 'imaginative' feelings of propositions (i.e. qualities illustrating specific regions) when they are integrated with the emotional background in the form of a contrast. In the case of such feelings there is a contrast between our awareness of a sense-quality characterizing a spatial region at a particular time, and the emotional background of activity, which is the way we directly experience the events in causal nature. We thus obtain the vivid conscious awareness, say, of red located in some particular region reinforced by this emotional background.

However, for Whitehead 'intellectual feelings' form only a narrow section of experience, as there are large tracts of experience which have not the peculiar vividness of consciousness. Thus we find Whitehead saying: 'This account agrees with the plain facts of our conscious experience. Consciousness flickers; and even at its brightest there is a small focal region of clear illumination, and a large penumbral region of experience which tells of intense experience in dim apprehension.'[2] Let us consider (*a*) first. For Whitehead a 'conscious perception' is a simplified type of intuitive judgment,[3] since it is one of the simplest forms of comparison (or contrast) between the perceived region illustrated by sensory qualities and our direct causal awareness of the physical events, which we take the former to symbolize. Such intuitive judgments may, as we have seen, lead us into error; for example, we may see a coloured patch and assume that it is an armchair, whereas it may only be a mirror image.

[1] P R, cf. p. 376 and pp. 382-9. [2] P R, p. 378. [3] P R, cf. p. 386.

As we have noted, in normal perception perceptual objects generally have associated with them certain 'conveyed' sense-qualities (contributed by the observer himself). In delusive perception there is usually a set of such conveyed sense-qualities which have, however, very little relevance to the events they illustrate. In such cases 'conscious perceptions' tend to approximate to 'intuitive judgments',[1] for they become 'imaginative notions about things'. There are, however, cases where 'intuitive judgments' tend to approximate to 'conscious perceptions'. For example, at breakfast we may suddenly close our eyes and picture the table we have just been looking at. In this case, the characteristics of the centrally excited bodily event which Whitehead assumes give rise to the image are very nearly the same as those due to the recently observed breakfast table. As we have noted, our attitude towards these two varieties of experience will differ. In perceptive feelings our attitude towards the perceived datum is one of unqualified acceptance, whereas in imaginative feelings our attitude (qualified acceptance) reflects the fact that the data have been imaginatively arrived at; in other words, we merely entertain them.

In the case of those 'intuitive judgments' which come under the heading of memory, the 'recollected qualities' resemble the characteristics of the past events, and the subject's attitude to them will have what Whitehead calls a belief character,[2] a belief that the 'recollected' qualitative pattern is similar in type to that observed in the past. He would probably say that it was just this close relevance to the previous occurrence which gave memory its characteristic fixity and order. Hume has pointed out that in memory the ideas concerned have a fixed order and position, as contrasted with the liberty of the imagination to transpose and change them. In the case of those 'intuitive judgments' which come under the heading of imagination (or fancy), the qualities originated by the subject may have very little relevance to past or present fact. Together with this, we find that the attitude of the subject towards the imagined data is usually one of 'inattention' (or indifference) to the truth.

Whitehead proceeds to discuss 'intuitive judgments' whose emotional pattern (i.e. subjective attitude towards the experienced datum) is dominated by indifference to truth and falsehood. This he calls 'conscious imagination' (i.e. imagination) in which 'We

[1] P R, cf. p. 384. [2] P R, cf. p. 378.

are feeling the actual world with the conscious imputation of imagined predicates, be they true or false'.[1] By this he means that whereas, for example, in remembering and reasoning there is attention to truth, in imagination there is inattention to it. When we lose ourselves in the fancies of our own creation there is a readiness to overlook the characteristics of actual fact. On the other hand, 'attention to the truth' is merely the refusal to allow one's critical faculty to be thus dulled.[2]

In imagination, then, little note is taken of the truth or falsity of the data entertained. There is no fixity or order, no adherence to the characteristics of some past event (as in the case of memory). There is the fluidity of imagination as contrasted with the pattern and order of memory.

We need next to consider intuitive judgments, in which there is 'attention to truth' (i.e. their truth character is taken into account). These can have three forms: (1) 'affirmative', (2) 'negative', and (3) 'suspended', which show themselves in the shape of different psychological attitudes on the part of the observer towards the experienced data.[3]

(1) 'Affirmative intuitive judgments' occur when the 'imagined pattern' is shown to be identical with 'the perceived characteristics of the events' to which it refers.[4] This is the case with memory. We may feel that the complex image we entertain resembles the characteristics of the previous occurrence—our experience will then contain the attitude of belief, as it does, for example, when we believe that William the Conqueror landed at Hastings in 1066, that Queen Anne is dead, or that the next door neighbour's cat is black. It should be noted that Whitehead is using 'belief' here to refer to our judgment of the compatibility (correspondence) of the complex image with actual fact; it is a feeling of the truth of the proposition. He is not dealing with 'belief' in the sense in which there is a measure of doubt due to an incomplete knowledge of the facts, as when, for example, one says 'I believe it will rain tomorrow', for the possibility is always open that it may not.

In the case of (2) or 'negative intuitive judgments', the imagined pattern conflicts with actual fact,[5] and this shows itself experientially as an attitude of disbelief.[6] We feel that the imagined qualitative pattern or proposition fails to harmonize with the rest of our experience, that it is not exemplified as such in our actual world.

[1] P R, cf. p. 388. [2] P R, cf. p. 389. [3] P R, cf. pp. 382-3 and p. 385.
[4] P R, cf. p. 382. [5] P R, cf. p. 382 and p. 387. [6] P R, cf. p. 385.

In other words, we are aware that the proposition is false. Thus we may disbelieve that there is cheese in the cupboard when we open it and find it empty. As Whitehead puts it, 'It is the feeling of absence, and it feels this absence as produced by the definite exclusiveness of what is really present'.[1] He concludes by pointing out that these 'two cases of intuitive judgment, namely the affirmative intuitive judgment and the negative intuitive judgment together with conscious perception, correspond to what Locke calls "knowledge" '.[2] In both these types of 'intuitive judgment' there is a feeling of certainty regarding the truth and falsity of the propositions with which they deal.

The more usual form of intuitive judgment, Whitehead tells us, is (3) 'suspended judgment', where the imagined pattern is shown to be neither wholly nor partially identical with the perceived fact.[3] In this case our attitude need not necessarily be one of belief or disbelief. By 'suspended judgments' Whitehead then seems here to be referring to what philosophers usually put under the heading of 'belief', as in 'belief' there is always a certain mental reservation, an element of doubt, for however firmly a proposition may be held, it may yet turn out to be false. Our beliefs may vary in degree ranging from an inclination to feel that way to almost certainty.

Thus for Whitehead 'suspended judgment' covers those cases where we are in some doubt as to the truth or falsity of the propositions concerned. It is through our ability to entertain such hypotheses that we are enabled to consider events which we are unable to verify directly. As Whitehead tells us, 'Our whole progress in scientific theory, and even in the subtility of direct observation, depends on the use of suspended judgments'.[4]

It may be of interest to compare Whitehead's view of 'belief' or 'suspended judgment' with that of Russell. Apparently, for Russell the whole of our intellectual life consists of beliefs[5] and of the passage from one belief to another by what is called reasoning. He analyses a belief into three elements: (i) the believing, (ii) what is believed, or the contents of our belief, and (iii) the objective, the events to which the contents refer. It is the objective reference of the belief that makes it true or false. Thus if we believe that Columbus crossed the Atlantic in 1492 the objective of the belief is the crossing of the Atlantic by Columbus in 1492.

[1] P R, p. 387. [2] P R, p. 387. [3] P R, cf. pp. 382-3.
[4] P R, p. 388. [5] *Analysis of Mind*, cf. p. 231.

But what is believed is not the actual fact which makes the belief true, in this case the voyage, but the present event in the mind related to the fact.

In the light of this, let us see how Whitehead would analyse the same belief-situation. According to him, it will consist of a (qualitative) pattern originated by the subject which has a reference to the actual voyage. This reference shows itself (in our experience) in the form of our awareness of certain data, acquired during early schooling, which have a complex relationship to these past events. Russell, however, tends to oversimplify the complexity of this objective reference when he speaks of the belief as pointing towards or away from the fact,[1] as if it were a compass needle.

Further Russell, at least in his *Analysis of Mind*, does not appear to distinguish independent attitudes of 'belief' and 'disbelief', for he tells us, 'We believe that some beliefs are true, and some false',[2] i.e. we have the same attitude of belief whether we judge the proposition to be true or false. Truth and falsity attain almost a Platonic character for him. For Whitehead, however, that which is believed or disbelieved is not merely the proposition, but its identity or incompatibility with the experienced data (which have this complex relationship). What we have therefore in the case of a judgment is the awareness of the truth relation (i.e. compatibility or incompatibility) existing between the imagined pattern and the experienced data, these two components being related together in the same experience.

II

Let us consider Whitehead's theory of judgment in more detail. He tells us it can equally well be described as (i) a correspondence theory, or (ii) a coherence theory;[3] and this seems to be largely bound up with the distinction he makes between a judgment and a proposition.

We learn that, whereas the judgment is purely a psychological attitude in the judging subject, the proposition which is one of the components in the experience is not restricted to that subject, for it 'can constitute the content of diverse judgments by diverse judging entities'.[4] Thus we can be aware of the truth of the proposition 'It snowed today', by remembering that such an event

[1] *Analysis of Mind*, cf. pp. 272-3. [2] Ibid, cf. p. 254.
[3] P R, cf. p. 269. [4] P R, p. 273.

actually occurred, and further verify it by looking out of the window and observing the snow thickly piled up on the garden path. But for another man placed under a blazing sky in the centre of the Sahara the same proposition would certainly be false.

Whitehead explains why his theory of judgment may be regarded as a correspondence theory. 'It is a correspondence theory, because it describes judgment as the subjective form of the integral prehension of the conformity, or of the non-conformity, of proposition and an objectified nexus'.[1] In other words, it describes judgment as the complex feeling (having the character of belief or disbelief) which arises from our awareness of the compatibility or incompatibility of the proposition and the experienced fact to which it refers, i.e. our awareness of a truth relation existing between these two components in our experience. But as we have noted, they can also be data for other judging subjects. Most of us can, for example, entertain the proposition 'Charles I had his head cut off', and compare it with our knowledge of this fact gleaned at some time or other from a history book. Thus the elements between which there is such a correspondence, unlike the actual feelings of them, can be considered by a variety of judging subjects at other places and times.

Whitehead goes on to tell us that, as this judgment is concerned with the conformity of two components within one experience, it may also be regarded as illustrating a coherence theory.[2] At first sight it might be thought that, as the relationship between the proposition and the experienced fact is one of conformity or nonconformity, they can also be said to cohere with each other. But Whitehead is referring rather to the coherence of the feelings (or awareness) of the two components in one experience into which the attitude of assent or dissent may enter. As he puts it, 'A judgment is a synthetic feeling, embracing two subordinate feelings in one unity of feeling'.[3]

For Whitehead, then, a proposition can be true or false, whereas a judgment can be correct, incorrect or suspended. He believes that there should be a correspondence theory of the truth and falsity of propositions, and a coherence theory of the correctness, incorrectness and suspension of judgment. This character of the judgment is dependent upon whether the feelings of the 'proposition' and the 'objective' in the judging subject harmonize or conflict with each other, and they in their turn depend on the

[1] P R, p. 269. [2] P R, cf. p. 270. [3] P R, p. 273.

conformity or nonconformity of the experienced elements. Hence, in accordance with the truth or falsity of the proposition, the coherence or incoherence of these feelings leads to the appearance of the appropriate belief attitude. In the case of a true proposition there is an awareness of the identity of the elements concerned which manifests itself in a feeling of harmony or correctness. On the other hand, in a false proposition the awareness of the diversity of these elements gives rise to a feeling of discord (or incorrectness) in the judging subject. Whitehead's point then is that in the case of the proposition we are only concerned with its truth-relation to the fact which makes it true or false. In judgment, however, we deal with the way this predication (or the content of the judgment) is felt by the subject, i.e. with our experience of its correctness or incorrectness.

Now it may be objected that Whitehead's use of the term 'coherence' here does not accord with its usual philosophical usage, namely that a proposition is true when it coheres with our general structure of beliefs (or propositions). It might therefore be said that Whitehead's theory of judgment is really a correspondence theory. This certainly seems to be the case. Nevertheless, we must not overlook the difference between what corresponds on Whitehead's view (i.e. the proposition and the objective) and what coheres, the prehensions of these feelings in the judgment. Though the term 'coherence' is somewhat misleading, in this context it does, however, bring out the dependency of the judgment, i.e. the feeling of correctness or incorrectness, upon the coherence or incoherence of these feelings in the judging subject.

Judgments for Whitehead are divisible into two species,[1] (*a*) intuitive judgments, (*b*) derivative judgments. As we have seen, intuitive judgments concern those cases where we verify the truth of a proposition by reference to the empirical facts. There are, however, other judgments occurring, for example, in logic and mathematics, where the truth of propositions is solely dependent upon logical tests of consistency.

This sort of judgment is dealt with by Whitehead under the heading of (*b*). He points out that, unlike intuitive judgments, where the proposition is predicated of empirical fact, in derivative judgments our attention is centred rather on the proposition itself, on its complex of qualities and relations; there is no attempt at empirical verification. It is because of this that they have to be

[1] P R, cf. p. 271.

guaranteed by logical criteria. In these judgments the attitude of assent or dissent will therefore depend upon whether or not the proposition stands up to such logical tests.

In derivative judgments we are then concerned with what is usually put under the heading of deductive reasoning. And unless we keep to certain logical rules error might easily creep in during the process of reasoning. Thus Whitehead tells us 'Logic is the analysis of the relationships between propositions in virtue of which derivative judgments will not introduce errors, other than those already attaching to the judgments on the premises'.[1] We may therefore say that the term 'derivative judgment' applies to cases of deductive reasoning where we deal with propositions and the relations of implication between them, their validity depending upon the test of self-consistency rather than empirical verification.

SUMMARY

Consciousness only appears in the more complex types of experience, which are put under the heading of 'intellectual feelings'. Two types are distinguished, (*a*) conscious perceptions, (*b*) intuitive judgments. These might be said to be analogous to sensations and images.

Whitehead distinguishes between (1) intuitive judgments whose emotional pattern is dominated by indifference to truth and falsehood, which he calls 'conscious imagination', and (2) intuitive judgments in which there is attention to truth. The latter has three forms; (*a*) affirmative, (*b*) negative, (*c*) suspended. In (*a*) the proposition is felt as being true, in (*b*) as false, and in (*c*) it is entertained as a hypothesis.

Whitehead's theory of judgment can equally well be described as a correspondence or as a coherence theory. It is a correspondence theory because he describes judgment as the complex feeling of belief or disbelief which arises from the compatibility or incompatibility of the proposition and the experienced fact to which it refers. It is a coherence theory since it is concerned with the coherence of our feelings of these two components in one experience.

Judgments are divided into (1) intuitive and (2) derivative. Intuitive judgments concern those cases where we verify the truth of a proposition by reference to the empirical facts. Derivative judgments deal with the judgments of logic and mathematics.

[1] P R, p. 272.

CHAPTER XI

TRANSMUTATION AND SENSE-PERCEPTION

I

IT is evident that Whitehead has to explain why on his account we perceive not a series of forms of energy implicated in events, having electronic or molecular characteristics (i.e. a patterned intertwining of nexūs), but tables, chairs, trees, etc., which have their own particular kind of existence. In other words, he is faced with the problem how, if, according to the physicist, what we have given to us is an atomic world, it comes about that we yet perceive it as made up of continuous regions passively illustrated by certain qualities. How does continuity arise from atomicity?

Whitehead stresses, for example, the difference between the paving stone as perceived visually and its molecular activities as described by the physicist. Pragmatically, he tells us, a paving stone is a hard, solid, static, irremovable fact. But this, however, is a very superficial account if physical science be correct. He therefore thinks that our sense-perceptions omit any discrimination of the fundamental activities within Nature, since the former are extraordinarily vague and confused modes of experience,[1] vague and confused because they mask and blur the atomic physical activities.

Whitehead, as we have seen, assumes that in sense-perception this change is brought about by a process of simplification (or transmutation), as a result of which the characters illustrating the many individual events are fused into one sensory quality. The sensory quality is then integrated with the spatial region, also derivative from these events, so that, for example, we see before us the passive material stone with its quality of colour. 'Mentality,' he therefore concludes, 'is an agent of simplification.'[2] And for this reason he thinks that our perceptual field is an immensely simplified edition of reality, namely, of the processes of activity

[1] N L, cf. pp. 64-5. [2] A I, p. 273.

or atomic events in the physical world. By 'mentality' in this connection he is referring to the process which simplifies these physical activities into the observed perceptual object.

He goes on to explain in more detail why if, according to the physical account the perceived object is a welter of physical activity, we yet perceive it as a passive substance illustrated by certain qualities. 'The association of a nexus as one with a quality will, for the experient subject, be in general a mode of exemplification which differs from the mode in which the respective individuals illustrate it.'[1] And he proceeds to give as an example the discipline of a regiment, which apparently inheres in the regiment in a different mode from the way it inheres in the individual soldiers. However, when Whitehead supposes that the difference between our perception of the passive inherence of qualities in regions, and the individual activities described by the physicist, is analogous to the difference between the discipline of a regiment and the discipline of the individual soldiers, he overlooks that these two cases are not parallel. The discipline of a regiment to which Whitehead refers characterizes a perceptual situation taken as a whole. It is not, then, epistemologically of the same type as the entities postulated by the physicist which have a higher degree of abstraction, and it also differs from our simple sensory qualities. We cannot single out the character of discipline in the same way as we can 'red'; at the most the former could only be a quality in the sense of a Gestalt quality. Further, though the discipline of a regiment may perhaps be shown to be a statistical average of the discipline of the individual soldiers, there seems no evidence to show that a similar relation exists between the colour and the electro-magnetic vibrations. The colour we perceive is certainly not a statistical average of some characteristic owned in common by the individual electrons, e.g. frequency of vibration. It is not like the white patch we see on the hillside, which on closer examination turns out to be a flock of sheep. White as a perceptual quality is not inherent in the electro-magnetic vibrations (or nervous impulses) in the same sort of way as it is in the case of the flock and the individual sheep.

Consider, for example, Whitehead's illustration of the way social groupings are perceived by us as single entities. 'It often happens,' he tells us, 'that in this perception of the nexus as thus qualified there is a wavering between the ascription of the quality

[1] A I, pp. 273-4.

to the group as one, and to its individual components as many. Thus, the orchestra is loud as one entity, and also in virtue of the perceived loudness of the individual members.'[1] Now it is patent that this is not the case with perceptual objects. When we are aware of a smell coming from a tube of hydrogen sulphide, we do not waver between ascribing it to the observed yellow liquid, or to the electrons out of which the atoms of hydrogen and sulphur are said to be composed. Further, though there is some qualitative similarity between the sounds made by the orchestra and the individual instruments, there is none between the appearance of H_2S in a test-tube and the individual physical activities. To assume that electrons have a smell, colour, or even an emotional tone, would be to forget that for the physicist they have only quantitative properties. But in fairness to Whitehead it must be pointed out that he is maintaining a realist philosophy of science, and that he does assume that electrons have a real existence. For example, his criticism in C N of the 'fallacy' of bifurcating nature into two spheres of reality commits him to the view that electrons have colour, etc., even if only in some attenuated emotional form.

Whitehead seems to think that for animals the most important example of transmutation is afforded by the physiological account of perception. No doctrine of sense-perception, he tells us, can neglect the teaching of physiology: 'the bodily organization is such as finally to promote a wholesale transmutation of sensa, inherited from antecedent bodily functionings, into characteristics of regions'.[2] He is arguing here that the forms of energy impinging upon the animal body from the external world undergo, when transmitted along the nerve routes, a process of modification or simplification, so that they finally appear in our perceptual field as clear-cut sensa in definite regions of space. But we should note that by the 'inherited sensa' he is really referring to the nervous impulses transmitted through our bodies, which, he assumes, are the primitive counterpart of the perceived sensa. When Whitehead therefore tells us that we must keep in mind the primary status of the sensa as qualifications of affective tone—'They are primarily inherited as such . . . and then by "transmutation" are objectively perceived as qualifications of regions'[3]—we need to remember that by 'qualifications of affective tone' he is not referring to anything given in conscious experience. He is referring to the nervous impulses which, during their transmission from event to event

[1] A I, p. 273. [2] A I, p. 275. [3] A I, pp. 314-5.

along the nerve routes, become differentiated into the various types of sensory qualities—sound, sight, touch, etc. To use his own terminology, 'causal efficacy' (in this case, the physiological transmission along the nerves) becomes transmuted into 'presentational immediacy' (or sensory perception).

Whitehead further amplifies this point when he states: 'In their most primitive form of functioning, a sensum is felt physically with emotional enjoyment . . . of its sheer redness'.[1] But apparently during the bodily transmission there is a gradual modification of these functions of sensa, so that finally they become transformed into affectively toned sensory qualities characterizing contemporary regions of space. As he puts it, 'for the initial occasions within the animal body, they are qualifications of emotion—types of energy, in the language of physics: in their final functioning for the high-grade experient occasion at the end of the route, they are qualities "inherent" in a presented, contemporary nexus'.[2] It is evident that the above account is not merely descriptive of our perceptual field, but also describes the physiological changes in our nervous system set up by external stimuli. The 'emotional enjoyment of a sensum of sheer redness' then refers in this context to the physical stimuli and the nervous excitations produced by them. And Whitehead proceeds to identify such scientific entities with 'Hume's "impression of sensation" stripped of all spatial relations'.[3]

We have another object lesson of how easily we may be misled if Whitehead's philosophical comparisons are taken at their face value. However, since Whitehead regards the physical and neural transmission of character as a more primitive form of perception, the above analogy, at least on his view, is not entirely misplaced. He therefore takes these 'emotional forms transmitted from occasion to occasion' as the primary facts of his theory of perception, just as impressions are in Hume's theory. But Whitehead would say that, unlike Hume, he was talking about sense-reception, the changes produced in our nervous system by physical stimuli, and not about sense-perception, the final clear-cut qualities in our perceptual field, which form Hume's primary data.

[1] P R, p. 446. [2] P R, p. 446. [3] P R, p. 446.

II

Lest the reader should suppose that transmutation is only a physiological and psychological process occurring during perception, we need to point out that Whitehead also thinks of it as operating in the physical world. In his account of structured societies, for example, he endeavours to show why as a matter of fact there is not just a mass of uncoordinated microscopic physical routes of events in the external world, but what we usually term material bodies to which Newton's Laws of Motion apply. There is not just an aggregate of millions upon millions of electrons, molecules and cells, each having its own characteristics, as they are masked by the one defining characteristic of the whole pattern (whether it be volume, mass, or motion). As a result of their co-ordination into one structured pattern, the microscopic societies are given a certain stability and flexibility which enables them to adapt themselves to changes in their environment. For example, molecules in an inorganic body have a certain stability which they could not have outside it. Or, to take a more complex example, individual cells would not live very long if detached from the main animal body. Hence, he concludes, 'this development of mentality is characteristic of the actual occasions which make up the structured societies which we know as "material bodies" '.[1] These comprise societies of various types of complexity—crystals, rocks, planets and suns. Whitehead then accounts for the substitution of one such structured pattern (or physical object) by a process of averaging (transmutation), which emphasizes the common features, whilst relegating the conflicting features to the background.

What may, however, seem surprising is Whitehead's claim that this process of averaging (or simplification), which in physics is concerned with the linking up of microscopic effects with macroscopic phenomena, is a primitive mental operation. This probably arises from his belief that in principle the process of transmuting the microscopic physical routes of events in the external world into the macroscopic physical objects resembles the process in our bodies whereby the nervous excitations are integrated into the perceived object in our perceptual field. When, however, Whitehead goes on to suggest that the development of mentality is characteristic of the events which make up the structured societies we term material objects, it becomes evident that by 'mentality' in

[1] P R, p. 141.

this connection he is certainly not referring to any psychological process.

Of course, *prima facie* Whitehead may seem to be justified, since he supposes that physical entities are made up of a similar kind of stuff to that of our experiences. Hence, the process of simplification which he assumes to be characteristic of the physiological perceptual process is also taken to be a characteristic of the complex structured societies we know of as material bodies. After all, he could argue, certain neural processes are simplified by the bodily functionings to give rise to our perception of perceptual objects; why should not the societies of events in the external world be simplified in a similar way, so as to take on the form of material objects?

At least his argument would seem to run on these lines, but we think that in drawing this analogy Whitehead has really started the other way round, from physics rather than physiology or psychology. A similar problem arises in physics, where an attempt is made to link up the microscopic electrons with the large-scale effects studied by Newtonian physics. Lorentz, for example, attempted to bridge this gap by the introduction of suitable mathematical machinery. With its aid he was enabled to demonstrate that the equations of electron theory could be transformed into the equations expressing the relations between the observable large-scale effects. The method employed is to take averages of the various magnitudes over regions which are large in comparison with the microscopic objects. As Whitehead puts it, 'In such equations even a differential element of volume is to be supposed to be sufficiently large to average out the diverse agitations of the molecules, and to register only the general unbalanced residuum which to ordinary observation is the motion of the body'.[1]

Lorentz then assumes that, if we form at each physical point (i.e. a small volume) in macroscopic matter the averages of the field intensities, the sudden changes and fluctuations of the microscopic events are smoothed out, and we are left with the continuous mean values which can be measured by our instruments. We thus arrive at the continuous regions of ordinary matter. The passive macroscopic object can in this way be considered as an aggregate of such small continuous volumes or physical points, each containing a vast number of electrons.

Whitehead's whole conception of how the many microscopic

[1] P N K, pp. 17-18.

routes of events are co-ordinated together into one structured society or material object, having a single defining characteristic, would then seem to be closely bound up with Lorentz's theory. But of course such a view, if taken as anything more than a mathematical device, gives the microscopic fields, together with the process of averaging, an existential import they do not deserve. It assumes that the human body and the physical objects around it arise as a result of a process which averages or smoothes out the microscopic entities into the familiar macroscopic objects. However, in physics no one has yet observed such a process at work. As used in Lorentz's theory, the operation of averaging seems to be purely a mathematical one. Whitehead therefore does not seem to be justified in taking the further step of assuming that something similar to this occurs in sense-perception.

III

But even if the reasons on which a theory is based are unsound, it may nevertheless sometimes fit the facts, or at least agree with speculations put forward by specialists in the field of study to which it is applied, and this curiously enough Whitehead's theory seems to do. His Lorentzian account of the way in which the nervous impulses are, as it were, smoothed out into the continuous sensory qualities fits in with the speculations of Adrian in physiology, since the general form of the problem is the same: namely, how do certain atomic elements acquire a passive continuous character? In physics we deal with electrons and ponderable matter; in physiology with nervous impulses and sensory qualities.

Speaking of the similarity between the nervous excitation and sensation, Adrian noted that the simplicity of the relation is at once very natural and very surprising. 'It means that . . . the mental correlate is a very close copy of the physical events in the sensory nerves. The only kind of distortion which takes place in the transference from body to mind (or in the parallelism of the bodily and mental events) is that the sensations rise and fall smoothly, whereas the nervous message consists of a series of discrete impulses with pauses in between. Somewhere on the way between the two there must be a smoothing process which converts the disconnected impulses into a change of much slower period.'[1] We can imagine, he goes on, that the impulses are conducted to

[1] *The Basis of Sensation*, pp. 117-8.

some part of the nervous apparatus of the brain where the nervous process rises and declines much more slowly than in the sensory fibre. There is therefore no need to look outside the central nervous system for the smoothing process, which integrates the series of impulses into a quasi-steady effect.

What is interesting is Adrian's assertion that 'the mental correlate is a very close copy of the physical events in the sensory nerves', which seems to be precisely what Whitehead has been saying above. But they differ in this respect: Whitehead assumes that this similarity is both qualitative and structural, whereas Adrian restricts it to certain structural properties of sensations and nervous excitations. (Whitehead might, of course, argue that ultimately this resolved itself into a difference in intensity of affective tone.) Adrian also postulates a process by means of which the discrete nervous impulses are smoothed out into the continuous sensory qualities. On Whitehead's view, the discrete impulses become the physical feelings, and the smoothing out process the process of transmutation at some critical node of the brain.

Adrian's account, however, involves a difficulty; after the 'smoothing out' process has occurred, we should still be left with nervous excitations, and not sensations. The only difference would be that in one case they would be discrete, in the other continuous. As we have already noted, sensations and nervous excitations seem to be on different planes of generality, and it would therefore require more than a 'smoothing out' process to transform one into the other. Adrian's simple relationship is then much more complex than appears at first sight.

Let us look at Adrian's account in more detail. According to him, the stimulus (e.g. the shining of a light) appears suddenly and remains at a constant value. But, though the stimulus remains constant, the excitatory process in the receptor declines gradually, owing to adaptation. As the excitatory process declines, the intervals between the impulses transmitted along the sensory fibres become longer and longer. In the actual sensation the rise and decline is a fairly close copy of the excitatory process in the receptor, except that it is 'smoothed out'. But what should be noted is that there is no simple correspondence between the physical stimulus on the one hand, and the excitatory process or the nervous impulse on the other. However, a most important feature of Whitehead's view is that he believes that there is such a correspondence between stimulus, nerve-excitation and sense-perception; that they are all different forms assumed by the same sensum during its transmission from the external world *via* our body to conscious experience.

It is because Whitehead holds such a view that he can state when criticizing Locke's theory of perception, which regards secondary qualities as mental additions to physical nature; 'But throughout the whole story, the sensa are participating in nature as much as anything else'.[1] In the physical stage, red, for example, is felt with the 'emotional enjoyment of its sheer redness',[2] and as a result of the bodily transmission it becomes the clear-cut sensory quality directly observed by us. Or, to take the case of smell, 'The experience starts as that smelly feeling, and is developed by mentality into the feeling of that smell'.[3]

One thing is clear; Whitehead is certainly assuming there is something similar to these perceived sensa in nature, whether it be an 'emotional enjoyment of sheer redness', or a 'smelly feeling'. He postulates that there is a direct correspondence between (i) the physical stimulus (which he takes to be a pattern of affective tone), (ii) the nervous excitation, and (iii) the perceived sensory quality. But, as we have seen above, the structure of (i) is dissimilar to that of (ii) and (iii). Though the nervous impulse has some structural affinity with the sensory quality, the structure possessed by both is quite unlike that of the stimulus. The correspondence is not then as simple as Whitehead believes.

IV

The view that physical events and sensory experience have a common structure is the basis of Whitehead's theory of symbolic reference, according to which our sensory perceptions stand as symbols for the activities in the external world. As he tells us, 'Symbolism from sense-presentation to physical bodies is the most natural and widespread of all symbolic modes'.[4] His whole notion of symbolic reference is then based on his belief that there are identical elements connecting human experience with the physical world. 'This fact, that "presentational immediacy" deals with the same datum as does "causal efficacy", gives the ultimate reason why there is a common "ground" for "symbolic reference".'[5]

[1] P R, p. 462. [2] P R, p. 446. [3] A I, p. 315.
[4] *Symbolism*, p. 5. [5] P R, p. 243.

However, though our sense-perceptions may symbolize the external world, there does not seem to be as much common ground between them as Whitehead claims. Whitehead has, of course, another string to his bow, since he assumes that we have direct awareness of the physical events in nature through perception in 'causal efficacy'. Nevertheless, he is still left with the problem of co-ordinating these two systems of perception.

In any case, any attempt to use the notion of structural correspondence between physical stimuli and sense-qualities to help out the causal theory of perception will not get us very far. For example, we find Russell, who advocates such an approach, saying, 'As to intrinsic character, we do not know enough about it in the physical world to have a right to say that it is very different from that of percepts; whilst as to structure we have reason to hold that it is similar in the stimulus and percept'.[1] However, if the scientist in dealing with the physical world is, as Russell has so often reiterated, only concerned with structure, the question of its intrinsic character is quite irrelevant; and further, there seems no reason to believe 'that the stimulus must possess whatever structure is possessed by the percept'.

We need to note that, according to Adrian, the actual quality of the sensation, i.e. whether it be a sound, colour or smell, depends upon the sensory path the impulses travel. Apart from this there is little to distinguish the message (or stream of impulses) coming from the different receptors, since they only differ in frequency. There is nothing in the physical account itself to explain why a physical stimulus should give rise to a sound, colour or smell. With a suitable apparatus or specially adapted sense-organs, it might give rise to any one of these. It seems then that Whitehead's illustrations have led him to overstate his case, since the 'sensa' as they manifest themselves in physical nature can give no indication as to what kind of sensory qualities they may be transformed into. Whitehead would probably agree with this criticism and say that, though these physical 'sensa' are to be regarded as forms of emotional intensity, they are, when compared with the observed sense-qualities, qualitatively undifferentiated. Their differentiation is due to the human organism and the Category of Transmutation.

Let us take a parting glance at the Category of Transmutation.

[1] *Analysis of Matter*, p. 400. Russell overlooks that simply to say that two systems have the same structure is trivial.

We may ask, is it really necessary? In physics it is a mathematical device, in physiology a mere guess. Does it then merit the attention Whitehead has given it? Whitehead, however, finds it necessary to bring in this category to bridge the gap between the atomic physical activities on the one hand, and the passive perceptual qualities on the other. But the need for importing such a category only arises as a result of our giving the physicist's abstract account of the world an existential import it does not merit. One is then faced with the problem of how these atomic activities are transformed into our continuous sense-perceptions.[1]

SUMMARY

Whitehead is faced with the problem how, if, according to the physicist, what we have given to us is a world of atomic events, we yet come to perceive it as made up of continuous regions illustrated by qualities. The function of the Category of Transmutation is to bridge the gap between the atomic physical activities and the passive perceptual qualities. Whitehead believes that the forms of energy impinging on the animal body from the external world undergo, when transmitted along the nerve routes, a process of simplification, and finally appear in our perceptual field as clear-cut sensa in definite regions of space.

However, Whitehead also conceives of transmutation as operating in the physical world, and is concerned to show how the microscopic societies of physical events are co-ordinated into macroscopic objects. He postulates a process of averaging which emphasizes common features and relegates conflicting ones to the background. This conception seems to resemble Lorentz's theory, that ponderable matter arises from the operation of averaging over the electrons.

[1] It was noted by A. O. Lovejoy (*Revolt Against Dualism*, pp. 156-89) that despite his protests against the bifurcation of nature, Whitehead in some of his earlier writings put forward a form of dualism.

CHAPTER XII

SENSORY PERSPECTIVES

I

WHITEHEAD in his earlier works asserted that we were aware of a common external nature which also contained within it such qualities as colours, sounds and smells. But when he starts to analyse this common nature (or immediate field of perception) it turns out that these qualities are bodily dependent. And this is at the root of the multi-termed relationship which, according to Whitehead, sense-qualities have to events. Whenever we say there is a colour, there is always an implicit reference to the sensorium (or percipient event) which is happening simultaneously with the event in which the colour is situated.

But what can Whitehead mean by saying as he does that the sense-qualities are projected (or ingress) into nature? If they were literally shot out of our heads, then not only would normal percepts be public property, but also delusive ones, and nature would then be populated by all sorts of queer objects. Whitehead would seem to mean by this that they are only apparent characters of events due to the idiosyncrasies of the observer, and which may vary according to his spatial position and bodily states.

Whitehead's description of the many-termed relation a sense-quality has to events can be formulated as follows: 'Sense-object O with the situation σ for the percipient event π with the active conditions $\gamma_1, \gamma_2, \ldots \gamma_n$'. In other words, the appearance of a sense-object in a certain situation is dependent upon the bodily event of the observer and certain preceding causal conditions. If we simply took O, σ and π as descriptive of certain directly given classes of elements in our perceptual field, then Whitehead's account would certainly be an acceptable one, and resemble the Gestalt account of perception. However, once we bring in γ we also give a causal explanation of the appearance of O in σ, and the causal conditions seem to be already in a different category from the immediately perceived elements.

Whitehead in his account would then seem to slur over two different questions: (1) what is immediately given in naïve experience, and (2) our explanation of its causation. It may be that as a direct fact of perception the perceived quality is multiply related to the elements immediately presented. But it is quite another matter to suppose that their appearance in events is due to a causal relationship. He would, of course, assert that we have some measure of direct awareness of the causal activities in nature—though in a rather vague sort of way—and that this is what he means when he speaks of our 'apprehension of events'.

Whitehead in his earlier work might be said to be putting forward a theory of the direct perception of external events, and a causal theory of our awareness of the sensory qualities which mark them out. Further, though he did distinguish between the 'apprehension' of an event and the 'recognition' of its sense-objects, he tended to put both kinds of awareness under the heading of sense-perception. In PR he seems to regularize his earlier position by bringing in the two perceptive modes, 'presentational immediacy' and 'causal efficacy'.

II

Whitehead in PR[1] makes the difference between the sensory perspective (or, as he now calls it, the presented locus) and the duration (or field of physical activities) more precise. Though we usually assume that the sensory perspective is closely related to the actual physical events now happening and tend therefore to identify them, they are, he claims, quite distinct. The observed perspective is dependent largely upon our bodily functionings, which, though usually excited by the antecedent physical events, may yet arise through abnormal conditioning events, such as drugs.

When we are aware of the sense-qualities as illustrating some specific external region, Whitehead claims that we also have direct awareness that they are obtained through the antecedent functionings of our sense-organs; we are aware of them as efficacious in perception. For example, when we look out of the window into the garden and see a tree, not only are we aware of the tree, but we also have the vague awareness that we see it with our eyes. Most philosophical accounts of perception, Whitehead

[1] PR, cf. Part IV, Chapters IV and V.

argues, delete the fact of the bodily functioning and concentrate on the immediately given coloured patches in our perceptual field.

He goes on to point out that the bodily parts perceived as causal in perception are perceived with greater distinctness by means of bodily sense-data. Examples of such sense-data are the slight eye-strain involved in seeing, or the kinaesthetic sensations in our arms when we push against obstacles. We need, however, to distinguish between these sensations, and the arm as an experienced unit bounded by certain visual data of shape and colour, in which the kinaesthetic sensations associated with effort appear. These kinaesthetic data and tactual sensations are still for Whitehead perceptions in the mode of 'presentational immediacy'. It is therefore not the case that such data and sensations belong to the sphere of 'causal efficacy', as has sometimes been assumed; perceptions in this latter mode have a more deep-rooted character.

Further, it should be noted that Whitehead's doctrine of 'causal efficacy' not only refers to our experience of bodily efficacy during the functioning of our sense-organs, but also to our awareness of the activities of the world around us. We are, he claims, directly aware of a surrounding efficacious world of related events: 'So far as concerns the causal efficacy of the world external to the human body, there is the most insistent perception of a circumambient efficacious world of beings. But exact discrimination of thing from thing, and of position from position, is extremely vague, almost negligible. The definite discrimination which in fact we do make, arises almost wholly by reason of symbolic reference from presentational immediacy.'[1]

Whitehead believes that the function of the perceived sense-qualities is to connect the immediately past events (which gave rise to them) with the events now happening, so that they may be said to come to act as symbols for them. In this way the perceived sense-qualities have some relevance to the contemporary happenings in nature. Thus when, for example, we cross a road and hear the sound of a horn, it can be said to warn us of the approaching motor car so that we move out of its way.[2]

[1] *Symbolism*, p. 65.

[2] One can already see in A E the beginnings of these two modes of perception. We imagine, Whitehead tells us, that we have immediate experience of a world of perfectly defined objects which is known to us by

III

The divergence between the field of sensory perception and the causal world of efficacy is best seen in Whitehead's doctrine of strains. By a 'strain' he means 'A feeling in which the forms exemplified in the datum concern geometrical, straight and flat loci'.[1] Whitehead is here referring to the spatial perspective illustrated by specific sense-qualities. The perceptual field (or focal region) is defined by the bodily or percipient event which is related to the rest of the events of nature co-present with it, and is to be regarded as an instantaneous snapshot of these happenings. The geometricized region (or spatialized duration) is then dependent upon the percipient. A 'strain-locus', Whitehead tells us, 'extends beyond that experient indefinitely, although defined by geometrical elements entirely within the extensive region which is the standpoint of the experient'.[2]

Whitehead distinguishes between the extensive relations of our sensory perspective and those of durations (i.e. the physical states of nature). He points out that whereas the perceived locus provides a uniform system of geometry pervading all regions, 'durations share in the deficiency of homology characteristic of the physical field which arises from the peculiarities of the actual events'.[3] What Whitehead is saying is that perceptual space, unlike physical space-time, has a systematic uniformity about it. In his earlier work he argued that space must be homoloidal to allow for congruence (identity of spatial judgments), as this was a necessary condition for measurement. He now seems to have modified his view so as to let physical space-time have the curvature required for it by Einstein's theory. But what he seems to overlook is that perceptual space itself is at the most only approximately uniform.

Whitehead, as we have already seen, believes that the appearance of sense-qualities in the perceptual continuum is determined by the physiological functioning of our bodies as well as by the antecedent physical occurrences. He goes on to argue that there is

the direct deliverance of our senses. But this world is a world of ideas, its internal relations are relations between abstract concepts. The fundamental question of scientific philosophy is then the elucidation of the precise connection between this world and the feelings of actual experience. (Cf. p. 158.)

[1] P R, p. 439. [2] P R, p. 459. [3] P R, p. 180.

also an isomorphic relation between these latter.[1] For example, when a pattern of light impinges on our retina, the frequency of the light and its geometrical (spatial) pattern are transformed into nervous impulses which are ultimately translated into sensory awareness. Now Whitehead claims that in the propagation of a nervous current along a nerve each such impulse has to be concerned with very nearly the same 'sense-datum' and 'geometrical pattern'. It may seem rather queer to talk of a nervous impulse as being characterized by a 'sense-datum'. But Whitehead is obviously referring to a form of energy having specific physical properties. This is once again an example of Whitehead's tendency to employ perceptual language when he is really talking about physical occurrences. As far as the 'geometrical pattern' is concerned, it seems a moot point whether it retains its identity, since it must undergo considerable distortion during the process of transmission; though of course this might be conceived after the fashion of some mathematical transformation of the original pattern. However, this does not prevent Whitehead from claiming that the perceptual end-product is a simplification (or an arithmetical average) of such a multiplicity of analogous sense-data and geometrical patterns. Or, as he puts it, 'a strain is a complex integration of simpler feelings'.[2]

Apparently during the process of perception our perceptual field is mapped out, as it were, by propositional feelings (i.e. regions qualified by sense-qualities). The changing event which is our body is given a position 'here-now'; the quality illustrating it may be just a feeling of bodily awareness. The focal region (or region where we say the perceived object is situated) is given a position 'out there', on which the sensa are said to be projected. The intermediary spatial regions between the observer and the perceived objects are mapped out in a similar way. Further, the antecedent nervous impulses through which the geometrical patterns and sense-qualities are transmitted also obtain some form of location and qualification, probably as states of bodily tension reinforcing the bodily awareness. In this way the whole of our sensory perspective is discriminated into regions illustrated by qualities.[3]

Whitehead, however, goes on to point out that in ordinary perception only one or two regions of our sensory perspective

[1] P R, cf. p. 442.
[2] P R, p. 439.
[3] P R, cf. p. 444.

are clearly distinguished. This comes about once again as a result of such a process of structural simplification (or transmutation). In this process emphasis may be put on the body as active in perception in which there is (i) 'the association of the sense-datum with the seats of some antecedent sets of feelers',[1] i.e. the characteristics of a group of antecedent bodily events are simplified so that we have an awareness of the body as a personal unity. On the other hand, the external region may be emphasized. In this case there is (ii) 'the association of the sense-datum with the focal region of the final percipient',[2] where, say, we experience a colour 'out there'. Finally, both the body and the external region as concerned with the same perception may be emphasized. We then have (iii) 'the association of the sense-datum both with the seats of antecedent feelers and with the focal region of the immediate feeler',[3] e.g. we may experience a feeling of red irritation as well as an awareness of red in front of us. As a result of this selective emphasis on specific regions of our perceptual field, we tend, Whitehead would argue, to neglect the great complexity of our perceptual functioning. We imagine that we directly observe concrete objects in the world around us which are independent of our perceptions.

Whitehead has been describing here in a very schematic way the mechanisms which he believes underlie our perception of the external world and our perception of bodily efficacy. In his earlier work he tended to take the world of perceptual objects as being largely a system of abstract concepts. In PR there has been a change in emphasis; the constructive process seems to have become a physiological rather than an intellectual one.

IV

In SMW, for example, Whitehead certainly implies that our immediate perceptions are merely private perspectives of some real objective world. This comes out in his assertion 'Thus if green be the sense-object in question green is not simply at *A* where it is being perceived, nor is it simply at *B* where it is perceived as located; but it is present at *A* with the mode of location in *B*'.[4]

In direct experience, however, the sense-object (or coloured

[1] PR, p. 445. [2] PR, p. 445. [3] PR, p. 445.
[4] SMW, p. 88.

patch) is located at the place where we see it. To say that it is also located or rather perceived in *A* (the body of the observer) is to relegate it to the level of a private sensation, to assume that though the colour is really inside our head we see it as located in some external region of space. Such a view presupposes straight away the initial privacy of our sensations and the causal theory of perception.

Now we do not deny that in perception we see beyond and away from ourselves. However, as soon as we start talking about perspectives, we assume that what we perceive (i.e. the perspective) is different from the physical thing really situated there. We also assume that the perspective stands in some form of relationship to the physical thing and the perceiver. Though one may say that in perception we are directly aware that we are related to other things in our perceptual field,[1] it is quite another matter to assume as Whitehead does that this perceived relatedness is really a perspective of certain entities in physical nature. This involves already a sophisticated process of interpretation. In naïve experience the thing and the perspective are one and the same. Though we are certainly aware that the thing over 'there' and the perceiver 'here' are spatially related to each other, there is no third term, the perspective, perceived in lieu of the thing, to which it stands in direct relationship.

When Whitehead speaks of every spatio-temporal standpoint as mirroring the universe, he asserts that if one thinks of it in terms of our naïve experience it is a mere transcript of the obvious facts: 'You are in a certain place perceiving things. Your perception takes place where you are, and is entirely dependent on

[1] This seems to have been roughly his view in P N K, pp. 13-14, where the two places or events *A* and *B* were parts of our perceptual field. Speaking e.g. of Alciphron's vision of the planet, he pointed out that what he directly knows is that 'I, Alciphron, am located in my percipient event "here and now", and the immediately perceived appearance of the planet is for me a characteristic of another event "there and now". In fact perceptual knowledge is always a knowledge of the relationship of the percipient event to something else in nature.' In S M W this 'immediately perceived appearance' takes on the definite character of 'the perspective of the thing over there from the standpoint here'. But as soon as we start talking about 'perspectives' we assume that what we perceive (i.e. the perspective) is different from the thing really situated there; that the perspective stands in some form of relationship to it and the perceiver. In direct experience, however, all that we have before us is the relatedness of our bodily experience to some other part of the field.

how your body is functioning.'[1] If, then, this perception conveys knowledge of the world stretching beyond us, it must be, he concludes, because the bodily life unifies in itself aspects of the universe.

But Whitehead's account is more than a mere transcript of naïve experience. He first talks about direct experience, that we are in a certain place perceiving things; he then goes on to tell us that this same experience is entirely dependent upon the way our body is functioning. This immediately presupposes the physiological account of perception, which is certainly not in fact a direct experience, unless, of course, one accepts Whitehead's theory of causal perception. There is no obvious connection between, say, our observation of the landscape stretching away in the distance and the light travelling to our retina from the objects in the physical world. The environment we observe extending beyond us is in our perceptual field and not in the world of physics.

As we have seen, Whitehead's way out of the difficulty is to assume that our immediate perceptual field is a perspective of the whole universe. He suggests that this perspective, though observed as stretching beyond us, is yet situated within our body, and that there is a continuity between body, mind and nature. Nevertheless, to say, as he does, that human experience is an 'act of self-origination including the whole of nature'[2] is certainly to go beyond the bounds of immediate sense-perception. We have little direct evidence for assuming that human experience is continuous with the physiological body and the natural world, that it is, as he suggests, a perspective of the physical world located within the body or brain.

Whitehead, as a result of postulating a continuity between mind, body and nature, accepts unreservedly the physiological theory of perception. The effect of physiology, he claims, 'was to put mind back into nature'.[3] One might be tempted to remark, remembering the experiments of Pavlov's school, that it has done the reverse, translating statements about direct experience into physiological terms. In any case, no physiologist has yet succeeded in tracing a continuity of transmission along our nerves to a projective reference of sensa beyond our body. From the point of view of direct experience, the physiological account is, to use his own phrase, 'a tissue of irrelevancies'. Even if one postulates a special mode of perception, as Whitehead does, there is still a

[1] S M W, p. 114. [2] A I, p. 290. [3] S M W, p. 184.

radical difference between our awareness of bodily functioning and the clear-cut account given by the physiologist in terms of nervous impulses.

The above discussion may be clarified if we note that there are at least two senses in which the word 'body' may be used: (1) as we immediately experience it in our perceptual field, and (2) as it enters into scientific descriptions, as a physiological organism. The perceived data are certainly beyond the body in sense (1) but not in sense (2). The whole notion of perceptual projection is an attempt to bring together in one system our perceptual data with the causal account of the physiologist, to explain why, if the thing seen is really inside our head, we yet perceive it as located in an external region of space. But as soon as we start talking about sensations located within our brain, we no longer talk physiology, where our account is restricted to nerve-excitations. We concern ourselves with a quasi-science, which is a mixture of physiology and psychology.

Whitehead believes that, since everything in nature implants aspects into everything else, we may have some direct emphatic experience of other people, for instance, of the anger in another person, and do not only have indirect knowledge obtained through observations of their overt behaviour. Thus, he tells us, 'we must allow for the possibility that we can detect in ourselves direct aspects of the mentalities of higher organisms'.[1] Now this contention seems plausible if one accepts Whitehead's postulate of the continuity of mind and nature, for other minds as well as our own are also within nature. But if one stresses the intrinsic differences between psychological experiences and physical events, it becomes difficult to see how there can be such a continuity of transmission.

Even if we accept the continuity postulate, our awareness of aspects of other people's minds would not be the same as direct acquaintance of them. It would be merely awareness of the effects set up by these other minds in physical nature. Thus even if we were acquainted with what was going on in, say, Jones's mind, so that we could follow his most intricate trains of reasoning and feel the same emotions as he does, we would merely be, as it were, using our body or mind as a recording instrument. Our knowledge of Jones's mind would still be as indirect as the knowledge we get of it from observation of his external bodily behaviour.

[1] S M W, p. 187.

SUMMARY

Sensory qualities on Whitehead's view of nature turn out to be bodily dependent, and this is at the root of the multi-termed relationship sensory qualities have to events. In perception we are aware of an interrelated spatial scheme whose various parts are differentiated by colours, sounds, etc. We also have direct awareness that these qualities are obtained through the antecedent functionings of our sense-organs, i.e. we are aware of our senses as efficacious in perception. The function of the perceived sense-qualities is to connect the immediately past events with the events now happening, so that they come to act as symbols for them.

Whitehead thus believes that there is a continuity between mind, body and nature. The whole notion of perceptual projection is an attempt to bring together in one system our perceptual data with the causal account of the physiologist.

CHAPTER XIII

PERCEPTUAL CAUSALITY, TIME AND TELEOLOGY

I

IT is of interest to see how Whitehead's account of the creative advance of nature works itself out in the field of conscious perception, which he puts under the heading of complex feelings. We find him criticizing the strained attitude of those philosophers who examine direct perceptual experience by acts of introspective analysis, which lift out the clear-cut sensory data and neglect the other characteristics of our experience, such as its aesthetic properties and its character of passage. It requires, he says, considerable ability to make the abstraction of bare sense-perception from the massive insistency of our total experiences. Our observed field is made up of related events, unit factors which have the character of passage about them, forming a background of activity against which the clear foreground of colours, sounds, etc., is displayed. We are not immediately aware of a succession of lifeless sensa, but of a passing flux of events; each event is seen to develop into the event which is future relatively to it—there is an observed continuity of transition between them.

For example, as we walk through the main street of a small market town and are jostled by the passing crowd, the changing scene develops, the costermonger pushes his barrow, and we move out of the way of the oncoming bus. The whole situation develops like the playing of a tune. The character which the later event assumes can be said to be determined by the event which has just gone before it. But by 'determined' here it should be understood that Whitehead is not offering an explanation in terms of disconnected events forcing each other, but merely describing a process of development. We take it for granted that the event which is the approaching bus will develop into the occurrence which is the bus passing us, and that it will not vanish into thin air, which it might well do if it were arbitrarily connected.

For all practical purposes, we never doubt that the events which are now happening conform to those in the immediate past. We observe this 'process of development' in the same direct way as we do the particular colours and sounds in the world around us, but it is a characteristic belonging rather to a whole situation than to any particular sense-quality. Our entire way of living, of making short-term forecasts, is dependent upon our awareness that we are acquainted with 'processes in development' rather than isolated events. When we see the earlier part of the process we are enabled to anticipate its later phase, and adjust our behaviour towards it accordingly; for example, we move speedily out of the way of the oncoming bus. We do not behave in this way merely because of some past association, or because of some elaborate argument by analogy.

For the Gestalt Psychologists too 'There is no mere sequence of indifferent events, connected indirectly. Each phase of what happens grows out of its predecessors, depending upon their concrete nature'.[1] Whitehead claims that he is describing what the naïve individual actually observes, that he is directly aware of such determination. He would say that he was merely pointing out the crude perceptual data from which the more refined notions of science are derived, in the same way as he points out the elements of our experience from which we obtain the notions of physical space and time. And, as in the latter case, there seems no reason why this awareness of direct determination, of the way in which one perceptual situation grows out of another, should not be some guide to the causal process in the physical world. Nevertheless, despite Whitehead's claim to have started from a naïve unbiassed approach to perception, he would seem, as we have already noted, to have based his descriptions of experience upon physical models. He might of course reply to this by saying that, as far as the passage of nature is concerned, both direct experience and physical nature have similar properties.

In our ordinary everyday life, Whitehead tells us, we become aware of what is usually called time as the succession of events; for example, that the state of affairs which was having our dinner came before the state of affairs which was our evening walk. But, as we have seen, this succession is not pure succession; each happening as it occurs develops into a new event, which conforms in character to the event which has just preceded it. Whitehead's

[1] Köhler, *Gestalt Psychology*, p. 390. (New York 1929).

view, then, is that the serial time talked about by such philosophers as Hume and Kant is merely an abstraction from the more concrete fundamental relationship of the conformation of the events in the present to those in the past. The sophisticated notion of time which they held, he argues, deletes the perceptual character of passage from the temporal process, transforming it into a succession of abstract units. In other words, it is a second-order concept.[1]

Whitehead believes that the denial by these philosophers of any direct causal determination stems from the oversimplified notion of time which they accepted. He therefore finds it necessary to distinguish between the crude perceptual character of passage which is fundamental, and the measurable serial time of science (and civilized life) which is a logical construct merely representing some of its properties. As he tells us, to attempt to dissociate time from the passage of events, and to suppose that it is anything else but an abstraction, is like the effort to find substance in a shadow. For him there is only time because there are happenings, apart from happenings there is nothing.

The problems which have arisen in philosophy as to the relation of causation and perception, Whitehead therefore thinks to be due to a faulty account of our immediate experience by such thinkers as Hume and Kant. According to them, the only data given to us in perception are a succession of sensa. And in order to explain why we are obstinately aware of a passage of events which causally conform to each other and not of such a mere succession, they found it necessary to introduce in one case a psychological habit, in the other a category of thought.

Whitehead now examines Hume's doctrine of causality.[2] According to Hume, he tells us, the appearance of certain antecedent percepts, for example a red hot poker, leads us to expect the consequent with which it has in the past been associated—a feeling of pain when we put our hand on it. This is all that causation comes to on such a view; it is a habit due to the frequent associations of certain past impressions. Any awareness of causation is read into the succession by us. Causality then becomes a peculiar fact about the way our minds work rather than about perceived nature.

Whitehead proceeds to test Hume's theory by applying his explanation to reflex action, for example the blinking of a man's eye when a light is suddenly shown in the dark.[3] He considers

[1] *Symbolism*, cf. p. 41. [2] P R, cf. p. 245. [3] P R, cf. pp. 245-6.

the subjective experiences of the man; restricting oneself to the sequence of sensa, they are—flash of light, feeling of eye-closure, instant of darkness. But the man will also assert if questioned, Whitehead claims, that he distinctly felt the flash of light give rise to the blink. It is this awareness of determination, that one experience has developed out of the other, which enables him to distinguish that the flash was prior to the blink. The whole occurrence hangs together in a connected pattern of experience and is not simply a sequence of sensa.

Whitehead's view is also shared by Gestalt psychologists like Köhler,[1] who tells us that on such a view as Hume's we might feel uneasy near a hot radiator, which would be one experience; then as another separate experience, we might feel an impulse away from it; and eventually as a third independent feeling we would feel ourselves moving away from the radiator. But this, he goes on, is an extremely artificial analysis. It is difficult to talk about these three experiences as if they did not belong together in one natural context; they are not indifferent experiences which might have been associated together in any other way.

Whitehead points out that Hume would intervene here, and deny that there is any direct perception of causal determination.[2] This he thinks is bound up (1) with Hume's *a-priori* dogma, that we are only aware of a succession of sensa, and (2) his attempt to reduce the man's feeling of compulsion to one of expectancy that the blink would follow the flash, i.e. to the feeling of a habit. But, Whitehead asks, how can a habit be felt when a cause cannot? Hume, then, seems to confuse a habit with our presumed feeling of it. One is a dispositional concept which on Hume's own test is a purely metaphysical chimera, since there is no impression of it; the other is our supposed awareness of this nonentity. To identify them, in any case, would be like identifying a general law with a particular instance of it.

Whitehead would then disagree with those philosophers who deny that we are directly aware of causal determination. In his earlier work, for example, in order to bring out the character of 'temporal passage' he distinguished between events and sense-objects. He was there aware of the shortcomings of the traditional theory of perception, which failed to take into account important properties of the perceived situation which could not be simply analyzed out in terms of sense-qualities. Whitehead conceives

[1] *Gestalt Psychology*, cf. p. 384. [2] PR, cf. p. 246.

'causal determination' to be rather a characteristic of a whole occurrence which extends over a period of time, than a specific sensation. It is only one of the many qualities in our perceptual fields which do not fit into the sensationalist scheme. As the Gestalt workers have indicated, our experiences are replete with such qualities, e.g. a circle is 'round', a decorative pattern 'symmetrical', etc. One might therefore, they say, just as well argue that there is no such thing as 'roundness' or 'symmetry', because there is no special impression of them.

Whitehead would agree that in a certain sense we do not see *A* cause *B*, if we mean by 'cause' here the kind of cause postulated by the scientist, or a relation of implication. (These seem rather to be limiting concepts obtained from the crude perceptual characters given in experience.) Causation, as the term is used in science or among philosophers who seem to seek the pure logic of it, is very different, he would say, from the crude causal determination directly observed by us. It is more general and abstract in character, and in the case of scientific objects has a much greater predictive value. It would certainly be wrong to suppose it could be a possible object of perception.

Koffka draws our attention to the possible connection between observed causal determination and physical causation, when he points out that behavioural causality may be a true clue as to the constitution of the real world. 'That would mean,' he goes on, 'that positivism has been too sceptical in its selections of behavioural data from which to build up a theory of the real world. . . . And we are not alone in claiming that possibly we have overshot the mark in our critical attitude. I need only mention Whitehead, philosopher and mathematician.'[1]

II

Whitehead's philosophy is sometimes said to be teleological in character. However, even a cursory examination will show that he uses the notion of teleology in a very different way from Aristotle. For Whitehead it is a description of the manner in which our experience develops out of the immediate situation rather than an explanation in terms of its essential nature.

When Whitehead says, for example, that an event moves towards its final cause (which is its subjective aim), he means that it has a

[1] *Principles of Gestalt Psychology*, p. 380.

'vector character', that it is a passage from the immediacy of the present into the future. When he tells us that the process of an event creating itself is dominated by a subjective aim which directs its process of realization, that 'This subjective aim is this subject itself determining its own self-creation',[1] he is merely drawing our attention to the 'perpetual transition of nature into novelty'.

If this is all that Whitehead means by 'teleology', why, it may be asked, does he apply this term to describe the development of an event? This is probably due to a misunderstanding on his part as to the way 'teleology' was used in Aristotelian dynamics. Apparently he seems to think that the essential difference between 'teleological' and 'causal' explanation lies in the fact that 'teleology' assumes a direction of events towards a goal. But this distinction is inadequate, as the causal explanations of modern physics use directed quantities (i.e. vectors). Since both the modern and Aristotelian views employ vectorial factors as a basis for their dynamics, there is on this point no difference between them. Hence, insofar as Whitehead chooses to describe the development of an event as 'teleological', because its passage from the past into the future can be described by definite vectors, there is at least here no divergence between his description of experience and the causal explanations of modern physics.

We need to indicate just what it is that differentiates the Aristotelian from the modern view. In Aristotelian dynamics the physical vectors are completely determined in advance by the objects concerned. For example, it was thought that the downward tendency of heavy bodies resided in these bodies themselves, that it was part of their 'essential nature' belonging to them once and for all. The situations in which they found themselves acted at the most as disturbing factors. In modern physics, however, we find that the existence of a physical vector is dependent on several variables. Whether an object will go up or down now depends also upon the kind of situation in which it finds itself. The essential difference between Aristotelian and modern dynamics is not that there is a direction towards a goal, but that in one case this direction is completely determined in advance by the nature of the object, while in the other it is a function of several variables, namely the object as in its relation to its environment.

Teleology conceived in its Aristotelian sense is therefore entirely opposed to Whitehead's position, since he emphasizes the

[1] P R, p. 96.

need for studying events in their functional relationships to each other. According to him the character which the initial phase of any event assumes is determined by the antecedent events, which in their turn depend on the much wider context of the rest of events in nature. Whitehead's notion of 'teleology' or 'subjective aim' is rather descriptive of the event's process of actualization. It says nothing about the event as striving to give rise to a specific character determined beforehand by its 'essential nature'.

Whitehead's reason for labelling the creative advance of nature into the future, 'final causation' is that it cannot be described as being due to an efficient or historic cause. Though (within limits) we can make an analysis of the present situation, showing that it has been derived from the past, this cannot be done with future events as they are not yet in being. One can only by analogy transfer this analysis to the present, and assume that the future is determined in terms of the present situation. What we really seem to be doing when we make long-range forecasts is to extrapolate the efficient (or historic) cause, based on the perceived conformation of the present to the past, beyond the present immediacy, in an attempt to limit the contingency of the future.

As a matter of experience we have found that the future does tend to conform to the present. Nevertheless, when we come to deal with it from the standpoint of our immediate experience, an element of contingency seems always to be present. In our everyday life we plan our actions to meet contingencies, as when going for a walk we take our umbrella with us in case it will rain. In attempting to understand past events we are faced, as it were, with a closed problem, as the events have already taken place, but when we come to deal with the future the problem still remains an open one.

Whitehead makes the above point in his ninth Categorial Obligation which he formulates as follows: 'The concrescence of each individual actual entity is internally determined and is externally free'.[1] When he says that the process of each event is 'internally determined', he is referring to the way the present situation has arisen from the events in the immediate past. 'But, on the other hand,' he goes on, 'the evolution of history is incapable of rationalization because it exhibits a selected flux of participating forms. No reason, internal to history, can be assigned why that flux of forms, rather than another flux, should have been illustrated.'[2] Whitehead seems here to be criticizing the view that

[1] P R, p. 37 and cf. pp. 63-4. [2] P R, p. 64.

there is some inner dialectic in nature which determines beforehand what type of pattern history will assume. The character the course of history assumes is purely contingent. There seems no inherent reason why that course of history rather than any of the many possible histories should have been actualized.

Nevertheless, in every pattern of history we can trace how certain events have arisen from events in their relative pasts, so that an entirely different course of history would also show this type of determination. As he indicates, the point from which we trace backwards this line of internal determination, and show how each event has developed from the other, is always in the past. However effective the concept of efficient causation may be when dealing with past experience, it seems to break down at the utmost point of the unfolding immediacy. This is what Whitehead wishes to draw our attention to when he says 'every instance of internal determination assumes *that* flux up to *that* point'.[1] In other words the particular course of events in which we trace such determination is already in the past, and is, as it were, laid out for our perusal.

The principle that the process of an event is internally determined and externally free may be summed up as follows. Each event can be said to be internally determined by the events in its immediate past, since they stand in finite spatio-temporal relations to each other (of successor and antecedent). On the other hand the specific character that event will assume in the immediate future remains in the sphere of contingency. Until an event has actually occurred it may exhibit any one of a number of alternative characters. But on the other hand, Whitehead would say that the future is not entirely contingent, since it must show some kind of internal determination. We ordinarily assume that it will be spatio-temporally connected to the present event, since by definition this is what we mean by the future—that it is relative to something going on now. Otherwise there would be little or no point in making predictions.

Bridgman is one of the few physicists who have had the courage to point out that the concept of time as generally held, especially its mathematical expression in scientific theorizing, fails to correspond exactly to that of experience. We all nearly always think of time, he points out, as a homogeneous and unlimited one-dimensional sequence, all past time on one side, all future time on the other, separated by the present which is in continuous

[1] P R, p. 64.

motion from the past to the future. This he claims is totally unlike the time of experience, which consists of 'a blurred sequence of memories culminating in the budding and unfolding present', having what he terms a unique apex with the possibility that everything may go awry. The Greeks, he goes on, must have differed from us in their feeling for time. We think of the future as stretching before us, and ourselves as going forward to meet it. Bridgman believes that this picture may be responsible for our feeling that the future has 'existence' and is essentially predictable. He contrasts this with the attitude of the Greek, who thought of himself as facing the past, with the future behind him coming up over his shoulder, as the landscape unfolds to one riding back to the engine in a train. Though, he adds, even this picture did not get rid of the idea of the 'existence' of the future; but it did emphasize that the future is unknown.[1]

Whitehead, however, unlike Bridgman, would not accept the view that the past completely breaks with the future. He brings to our notice that in broad outline certain general states recur, and that our very natures have adapted themselves to such repetitions, so that we instinctively believe in an order in nature. But he is careful to note that there is a complementary fact which is equally obvious; nothing ever really recurs in exact detail, no two days are alike, no two winters. The practical philosophy of mankind has thus been to expect the broad recurrences, and to accept the details as being in some measure unpredictable. As he puts it, 'Men expected the sun to rise, but the wind bloweth where it listeth'. Whitehead then makes room for these two contrasted notions in his account of nature—repetition and novelty. Nature is periodic, one day is followed by another, and every morning is a new fact followed by its measure of change.

SUMMARY

For Whitehead the field of conscious perception is made up of related events—unit factors which have the character of passage—against which the clear foreground of sensa is displayed. Our entire way of living, of making short-term forecasts, is dependent upon our awareness that the events which are now happening will conform to those of the immediate past. As a result of regarding perception merely as a succession of sensa, philosophers have

[1] Cf. Bridgman, *Nature of Physical Theory*, pp. 29-32.

assumed that causal determination is something extrinsic to the perceived event, instead of being a characteristic of it.

Whitehead's conception of 'teleology' appears to be consistent with modern physics. The essential difference between (*a*) Aristotelian dynamics, and (*b*) modern dynamics, is that in (*a*) the physical vectors are completely determined in advance by the object, whilst in (*b*) they are also dependent upon the situation in which the object finds itself.

Whitehead emphasizes in his account that not only is nature pervaded by definite patterns of order, but there is also always present an element of contingency.

CHAPTER XIV

SOCIETIES

I

WHITEHEAD'S concept of societies has a prominent place in his system and is given a highly generalized meaning, so that it is applicable to any set of related events. In its simplest form a society may be equated with a historical route pervaded by a specific character, i.e. an enduring object. As he tells us, 'An ordinary physical object, which has temporal endurance, is a society'.[1] For Whitehead the things which endure, the recognizable permanences in nature, tables, chairs, animals, etc., are societies, since he considers them to be historical routes of events, or strands of transition pervaded by a self-identical pattern. It is this 'social order' (or pervasion of the route by the pattern) which gives the object its peculiar character of permanence, which makes us say, for example, when we see the apple tree in the garden, that we are observing the same tree today as we did yesterday. However, it also has characteristics which vary with altering circumstances. In winter it is gaunt and bare, in spring it buds, blossoms and grows leaves, whilst in autumn fruit appears which ripens and falls to the ground. Unlike a particular event, then, such as a flash of lightning which happens and passes, a society or an enduring object has a history expressing its changing reactions to its changing surroundings.

It may be as well to glance at the way Whitehead defines social order. He tells us that a nexus enjoys social order 'where (i) there is a common element of form illustrated in the definiteness of each of its included actual entities, and (ii) this common element of form arises in each member of the nexus by reason of the conditions imposed upon it by its prehensions of some other members of the nexus, and (iii) these prehensions impose that condition of reproduction by

[1] P R, p. 47.

reason of their inclusion of positive feelings of that common form'.[1]

In simpler language, a historical route (or nexus) has what he terms a social order when (1) there is an identical sensory or physical pattern characterizing every event making up the route, (2) this pattern arises in each event by reason of its being inherited from other (antecedent) events in the same route, to whose character it therefore has to conform, and (3) as the result of such a transmission of a self-identical character from event to event, there is a continuation (or permanence) of character. Whitehead terms this self-identical pattern inherited throughout the route the defining characteristic of the society (or enduring object). This view that an enduring object is such a pervaded route is not new in his philosophy. For example, we find Whitehead pointing out in C N that all you mean by stating that Cleopatra's Needle is situated on the Embankment is that, among the structure of events which forms the medium within which the daily life of Londoners is passed, 'you know of a certain continuous limited stream of events, such that any chunk of that stream . . . has the character of being the situation of Cleopatra's Needle'.[2] It will be seen that Cleopatra's Needle defined in the above fashion approximates to the definition of a society given by him in P R. The 'continuous limited stream of events' corresponds to the 'nexus', what he terms a chunk of this stream to a 'member of the nexus'. The permanent object Cleopatra's Needle may be identified with the common element of form illustrated in each of the actual entities making up that nexus. Whitehead's reason for calling such a self-identical object a society is that he regards it as a set of related events, bound together, as it were, by a similar defining characteristic (i.e. complex eternal object or element of form).

It follows from this that whilst an object is in existence it is always adding to itself with the creative advance into the future. As he points out, a man adds another day to his life, the earth another millenium to the period of its existence. However, until

[1] P R, p. 46. Glossary of Terms:
nexus = historical route.
social order = reiteration of a specific pattern throughout the route.
common element of form = sensory or physical pattern.
prehension = the way each event inherits (or transmits) its character from (or to) some other event in the route.

[2] C N, p. 167.

the death of the man and the destruction of the earth, there is no completely determinate historical route which can unqualifiedly be said to be either the man or the earth.[1] When, for example, we speak of the immediate object in our perceptual field, we usually refer to the realized society up to that present stage of its existence. We may spill a bottle of ink on the carpet tomorrow, we may cut ourselves whilst shaving, or a bomb may fall on the house, so that things may take on quite a different look tomorrow. The characteristics pervading these respective routes therefore undergo within certain limits a progressive modification, though as in our last example it may sometimes be a very violent one.

Since any route of events pervaded by some defining characteristic (no matter how complex) is a society, Whitehead brings in the notion of personal order to distinguish those societies having a simple self-identity (i.e. not made up of simple objects or parts) from those which have a complex structure. A complex society is a patterned network of various strands of transition (routes of events) having diverse defining characteristics. An example of a complex society is an animal body made up of a multiplicity of cells each of which may be regarded as such a four-dimensional thread.

Whitehead gives two examples of societies having a personal order; they are (1) the life of a man, and (2) that of an electron. He amplifies this somewhat by saying that in a relatively simplified case an enduring object has personal order; it is a route which is purely temporal and continuous. A man defined as an enduring percipient, having a complex character in virtue of which he is considered to be the same enduring person from birth to death, is such a society. Whitehead is here referring to the stream of personal experience or the self which we discriminate as being localized within the region making up our body.

Apparently the most important characteristic of a society with personal order is that it is purely a thread of temporal transition. However, we do not agree with Whitehead when he assumes it has no spatial extension. Even the route of events making up the enduring percipient has some kind of localization. What we term the self (as Hume pointed out), at least as we notice it on inspection, is a pattern of feelings, wishes, desires, thoughts, memories, aches and pains, crowding in upon each other. As Whitehead himself points out, 'there are an indefinite set of obscure bodily feelings which form a background of feeling with items occasionally flashing

[1] A I, cf. p. 262.

into prominence'.[1] But, we may well ask, how can feelings form a background unless they are in some way spatially extended and localized?

The other example given by Whitehead of personal order, the life of an electron, seems largely a matter of interpretation. Taken as a particle, it certainly has extension, having among other things mass. On the other hand, if it is considered as a probability distribution, though it may lose what are termed its material properties, for example that of simple location, it then becomes anything but simple.

II

Whitehead proceeds to discuss 'corpuscular societies', i.e. physical objects (or historical routes) capable of being sub-divided into many strands of enduring objects: the tables, chairs, and other macroscopic objects of our everyday life which are analysable into molecules, electrons, etc. Each electron, molecule or cell within such a complex object is itself to be thought of as a society (i.e. a pervaded route of events). In the case of an animal body, we have groups of events patterned by electronic, molecular and cellular characteristics, which are co-ordinated within the wider pattern. It will readily be seen that the way Whitehead uses the term 'society' here corresponds to his use of the notion of 'organism' in S M W.

Further, as the individual characteristics of the included objects become more important in comparison with those of the whole pattern, so does that society become more 'corpuscular' in character until the notion of a defining characteristic has to be extended to cover the co-ordination of societies. To take an example given by him, a regiment is more of a 'corpuscular' society than, say, a man, since its defining characteristic is much more attenuated, referring as it does to a peculiar kind of social organization, rather than to a fixed set of qualities. We can more readily take note of its component elements, the men who make it up, and who are co-ordinated in a certain social pattern—the regiment—than we can the cells, molecules, etc., making up the man.

Whitehead amplifies the above by bringing in the notion of 'structured societies'. A 'structured society', he tells us, includes subordinate societies or routes of events functioning together in

[1] N L, p. 63.

one complex pattern. Material objects, for example molecules, crystals, and rocks, belong to the lowest grade of such societies. Animals and other living societies belong to the higher grades, the latter tending to be less stable, having by comparison a fairly short life-span.

By a 'structured society' Whitehead then means a co-ordinated group of such historical routes or microscopic sub-societies forming a macroscopic pattern having a specific defining characteristic. In the case of an animal body, some of its included routes may be of higher grades than others; for example, living cells are taken to be of a higher grade than inorganic molecules. As the defining characteristic of the whole society then increases in importance, so does that society take on a more definite structure. A table, for instance, has a more definite structure than a cloud or a pool of water.

Hence, as the macroscopic character of the whole society becomes more explicit, so does that society become more 'structured', overwhelming, as it were, the microscopic character of its included sub-societies. Thus, the picture of the Universe that Whitehead gives us is not worded in terms of material objects, but in terms of historical routes of events. We have a four-dimensional interconnected system in which are included subsidiary structures, the enduring objects, which can be further sub-divided.

As we have seen in the case of an animal body, some of the included sub-societies are of a markedly higher type than others. Whitehead tells us that in itself the animal body provides the environment which sustains the sub-societies of cells, molecules, etc., as well as the stream of experience which he calls the 'regnant' nexus, since this nexus has some measure of control over the functionings of the whole body.

However, only some of the routes in a living society are such that the component events exhibit any originality of reaction. Whitehead claims that it is precisely this feature which distinguishes a living from an inorganic route of events. The latter merely follows the ordinary causal laws of physics and lacks any element of novelty. In a 'living' society such routes having an originality of reaction are dominant, and in some measure exercise control over the structured society of events making up the animal body. To take the case of our own body, it is to some extent dominated by our stream of personal experiences, desires, volitions, and cognitions, which prevent us from being a mere slave to passing circumstances. We thus behave very differently from the thousand and one inanimate objects in our environment.

Whitehead distinguishes between inorganic and living routes (or nexūs) of events, pointing out that the former are societies in their own right. In other words, inorganic entities can have a separate existence apart from the particular animal body in which they may be included. We could, at least theoretically, trace the adventures of a suitably labelled molecule from the external world into our bodies and from our bodies out into the world again. But when we come to consider 'entirely living' nexūs (or streams of experience) we see that they stand in a rather different relationship to the animal body than do the molecules. Whitehead therefore thinks that we cannot have a 'route of experience' unless it is sheltered by some such animal body.

He now discusses the manner in which the 'stream of experience' is sustained by the animal body. 'A complex inorganic system of interaction is built up for the protection of the "entirely living" nexūs, and the originative actions of the living elements are protective of the whole system. On the other hand, the reactions of the whole system provide the intimate environment required by the "entirely living" nexūs. We do not know of any living society devoid of its subservient apparatus of inorganic societies.'[1]

The reader may feel slightly puzzled as to what exactly Whitehead means by such an inorganic system and its connection with experience. It is certainly peculiar to call the interconnected sets of elements making up the bodily cells an inorganic apparatus. In such a combination they form rather a biological unit. This is a consequence of Whitehead's belief that a living body consists of (1) inorganic entities such as electrons and molecules, and (2) a certain 'novelty of character' arising as a result of their combination in a specific pattern. He would here appear to be trying to explain why, if the body is composed of such inorganic elements, it has yet the character of life about it. However, this view overlooks that we only arrive at such inorganic entities after a process of analysis, when we destroy the biological unit and are left with something quite different on our hands.

We are given some indication of the sort of thing he had in mind in the above passage when he says, ' "Physical Physiology" deals with the subservient inorganic apparatus'.[2] In S M W, for example, when speaking of the way the cell effects changes in the

[1] P R, pp. 143-4. [2] P R, p. 144.

included molecules, he tells us that the question for physiology is the question of the physics of molecules in cells of different characters.[1] It would therefore appear that by 'Physical Physiology' he simply means physiology. The function apparently of this inorganic system is to protect the stream of experiences, and the originative actions of these elements—our volitions and cognitions—are in their turn protective of the whole body. By their aid an individual can adapt himself to sudden changes in his environment, and survive under the most adverse conditions.

III

Whitehead next discusses what he terms 'Psychological Physiology'. He is not, however, laying down the basic principles of some new science, but he is really talking about psychology. He tells us that it seeks to deal with 'entirely living' nexūs, 'partly in abstraction from the inorganic apparatus', for example, when we consider our experience in introspection, and 'partly in respect to their response to the inorganic apparatus', i.e. when we study the effect of the physiological functionings upon our sense-perceptions and emotions. It also attempts to deal with our experiences 'partly in regard to their response to each other', i.e. the way one experience gives rise to other experiences—feelings to thoughts, thoughts to desires and memories to memories.[2]

Whitehead now makes what seems to be a very questionable statement: 'It must be remembered', he tells us, 'that an integral living society, as we know it, not only includes the subservient inorganic apparatus, but also includes many living nexūs, at least one for each "cell" '.[3] It is certainly startling to learn that each group of molecules and electrons making up a cell has included a living nexus, which gives it its peculiar character of 'life', causing it to behave differently from inorganic entities. What he is trying to say, however, is that the cell as thus constituted has an originality of reaction not explicable in physico-chemical terms. Nevertheless, one does feel unhappy about Whitehead's mode of explaining such 'originality of reaction'.

Whitehead proceeds to conjecture some fundamental principles of Psychological Physiology. He considers a single cell, and he asks (1) whether the individual living occasions in abstraction from the inorganic events of the cell have a separate existence (a

[1] S M W, cf. p. 186. [2] P R, cf. p. 144. [3] P R, p. 144.

point we have already discussed), and (2) whether such a cell has included within it a mind or a stream of personal experiences having an identity of character.

Whitehead believes that, though the evidence at our disposal is extremely slight, it suggests a negative answer to both these questions. A living cell, he asserts, gives no evidence of such a unified mentality, it merely manifests a certain originality in its response to external stimuli. Life for Whitehead is therefore the origination of a certain novelty of character. That is to say, an organism is alive when its behaviour cannot simply be explained in terms of the antecedent physical events. This seems fairly obvious, since physics is not concerned with explaining the reactions of living organisms; this is primarily the job of the biological sciences. Physics only concerns itself with those aspects of nature to which the notion of physical causation is applicable.

When Whitehead speaks about life he takes up a curious dual attitude. It is evident that by an 'entirely living' nexus he means some type of mental experience or novelty of character derived from what he calls the 'inorganic apparatus'. On the other hand, when he speaks about originality of reaction, he is referring to the peculiar way in which the cell reacts to inorganic stimuli. Whitehead, however, does not clearly distinguish between them, and uses these different notions interchangeably. To say that life is 'originality of reaction' is not the same as saying that it is the introduction of conceptual novelty. In the former case he is speaking about the sort of data dealt with by the physiologist (or behaviourist) who finds that the cell has certain behavioural properties absent from inorganic entities. In the latter, what we are really dealing with is a philosophical attempt to explain, by the introduction of a doctrine of novelty, how mental experiences can arise in the animal body as a result of changes set up in the inorganic apparatus. But what needs stressing is that this whole account is purely hypothetical in character.

Though most people would agree that there is a radical difference between the stimulus and the observed reaction of the cell, not everyone would feel happy with Whitehead's attempt to outline the process which gives rise to this reaction. When he therefore says life is originality of reaction, he is not saying anything very new as far as physiology is concerned. But when he describes it as the emergence of novelty, meaning by this a 'living nexus' derived from the inorganic apparatus, then his position becomes questionable. In one case, he is giving a straightforward description of the cell's behaviour, in the other an explanation of it.

Whitehead then supposes that a complex structure of inorganic events is woven together to give rise to the route of events which forms our personal experience. As he puts it, 'The whole body is organized, so that a general coordination of mentality is finally poured into the successive occasions of this "personal society" '.[1] On Whitehead's view, mental experiences do not merely arise from each other according to their own laws, but are in the main derived from the antecedent functionings of the body.

Whitehead makes still another curious statement, when he tells us that the conclusion to be drawn from this argument 'is that life is a characteristic of "empty space". . . . Life lurks in the interstices of each living cell, and in the interstices of the brain. In the history of a living society, its more vivid manifestations wander to whatever quarter is receiving from the animal body an enormous variety of physical experience.'[2]

Though we may perhaps sit back and wonder when Whitehead says 'Life is a characteristic of "empty space" ', we need to take this statement at its face value. He supposes that these experiences wander through the interstices of the cell to some portion of the brain which is receiving, as it were, a torrent of experiences from all the individual cells, just as messages flood in upon a central telephone exchange from the outlying suburbs. The result of this transmission is that the functioning of the animal body does not entirely conform to the physical laws holding for the inorganic entities. A man does not behave in the same way as a stone of a similar shape and size. Unlike the latter, he can adapt himself to changing circumstances and acquire techniques which will enable him to obtain some control over his environment.

Instead of starting his account, as a biologist would, from the cell as a basic concept, Whitehead starts very much as a physicist might, from the inorganic apparatus. He is therefore faced with the problem of explaining why the cell as a unit behaves very differently from these inorganic entities. This, he assumes, is due to the generated field or living nexus which affects the inorganic apparatus in such a way that certain changes take place, which would not otherwise occur.

[1] A I, p. 271. [2] P R, p. 147.

IV

Though a route of experience or a living nexus cannot sustain itself apart from the animal body, Whitehead thinks, as we have seen, that it may be so canalized as to support a thread of personal order, or stream of conscious personal experiences. Such a conscious personality, however, only seems to be characteristic of the higher animals. We certainly have no ground for conjecturing living personality in the lower forms of life. Though he is of the opinion that the lower animals and plants are not dominated by any such experiences, he believes that the life arising from the individual cells within these societies is canalized into some faint form of social order. So that, for example, in a tree where there is a lack of differentiation of function we get something like a democratic control.

Whitehead refuses to accept an independent mind or soul presiding over the events making up the body. All the life in the body, he asserts, is the life of the individual cells. As there are millions upon millions of centres of life in each animal body, what he thinks needs explaining is the unifying control by means of which we have (*a*) unified behaviour, and (*b*) consciousness of a unified experience.[1] He believes that, among the higher animals, the events making up the living body are so co-ordinated that a stream of experience flows from the various parts into the brain, where there is produced the presiding or conscious personality. In their turn the emergent stream of volitions, cognitions, etc., modify the events throughout the rest of the body and give rise to its unified behaviour.

The above account is, however, beset with pitfalls. It is difficult to understand how a combination of inorganic entities can give rise to living experience. It is also difficult to understand how this mass of experiences flowing through the physical brain is co-ordinated to become our conscious personality, which in its turn sets up changes throughout the physical body, no doubt by acting on the nervous system.

This seems to involve a confusion; what Whitehead is trying to do is to show how the body described by the physiologist can be controlled by our direct experience. But in the physiological account these experiences are cut out altogether. What may be dealt with is, perhaps, electro-chemical changes in the cells of the

[1] P R, cf. p. 151.

cerebral cortex, which then transmit their influence through the nervous system, causing certain forms of behaviour to occur.

When Whitehead says the 'presiding occasion' is the final node or intersection of a complex structure of many enduring objects (i.e. the animal body), he no doubt means by this that the experiences derived from the individual cells in the body are finally poured into an interstitial space somewhere in the brain. By not clearly distinguishing between the personally experienced body and that described by the physiologist, Whitehead argues from one to the other. He hence puts not only our personal experiences, but also our experiences of the everyday world, within the physiological description. Whitehead also assumes that the animal body is made up of routes of direct experience and scientific events intertwined together in one pattern. He overlooks, however, that their defining characteristics are of different grades of generality. To talk of a 'subservient society' and a 'regnant nexus' may perhaps have some significance when we consider the body from a purely biological point of view. One is then concerned with similar types of things. But when Whitehead supposes that direct experiences, cells and molecules are co-ordinated together into the unity of one field (or structured society), then he is plainly mixing up entities of different grades of generality, physical and biological concepts with direct experience. And it is this confusion which we think is at the root of many of the paradoxes found in the pages of *Process and Reality*.

SUMMARY

The notion of a society is given a highly generalized meaning by Whitehead, since it covers among other things our everyday world of common-sense objects. A society or an enduring object has a history expressing its changing reactions to its surroundings. The notion of personal order is brought in by Whitehead to distinguish societies having a simple self-identity from those having a more complex structure. By 'personal order' he refers to an enduring object having only temporal continuity, e.g. the life of man.

Physical objects which are capable of being sub-divided into smaller parts are called corpuscular (or structured) societies. Inorganic objects belong to the lower grades of structured societies, animals and plants to the higher. Living societies, unlike inorganic ones, show an originality or novelty of reaction. In man a complex structure of inorganic events is woven together to give rise to our stream of personal experience. Conscious personality, however, only seems to be a characteristic of the higher animals. Nevertheless, Whitehead believes that the life of lower animals and plants is canalized into some faint form of social order.

CHAPTER XV

PHYSICAL THEORY AND THE PHILOSOPHY OF ORGANISM

I

THESE discussions of perception have shown certain difficulties which arise when interpreting Whitehead's work. One has always to be careful to distinguish those contexts in which he speaks of physical phenomena from those in which he refers to psychological experience, as he tends to use the same sort of epistemological language to describe them both. And indeed he attempts to set up an isomorphism between them.

However, though Whitehead's description of direct experience is in some respects more complete and satisfactory than that of other philosophers, yet his endeavour to draw a parallelism between the notions of physics and direct experience, especially in P R, cannot be justified. The point we have just made may be disputed, since it is usually believed that what Whitehead is concerned with in his philosophy is re-evaluating some of the basic notions of modern physical science. But this overlooks what Whitehead was doing in his earlier works, when he spoke, for example, of the 'fallacy of misplaced concreteness'; he was not criticizing the concepts of modern physics, but rather those of the Newtonian classical scheme from the standpoint of modern physical science. Indeed, he seems in P R to look upon the fact that he can draw a parallelism between modern physics and direct experience as an argument in favour of his philosophy.

It might, however, be argued that Whitehead's postulation of two types of causal process conflicts with the notions of physics, since in the usual type of deterministic physical system the historic (or efficient) cause of a process is not regarded as determinative of its dynamics. As Whitehead himself noted, even common-sense assumes that the greater number of antecedent events in the physical universe are irrelevant to the production of any particular effect. He tells us this assumption has grown out of the experience of mankind, as we usually take remoteness of time and place as evidence of comparative disconnection of influence.

But, insofar as the billard-ball conception of scientific objects has been replaced by the field theory, this view of physics (and common-sense), that in the consideration of causes the past history can to a large extent be neglected, breaks down. As Bridgman points out when speaking of electrical systems with radiation, 'such systems are determined by the present position and velocities of the charges in the immediate vicinity, and by the corresponding data at remote points given for proper epochs in the past; in this case, therefore, past and present history are necessary to determine the future'.[1] We have, at least here, some indication as to what physical notions would correspond to the dual causation put forward in PR.

Whitehead, for example, told us in AE when considering scientific objects as fields of force, that the field at any time due to an electron as a whole depends on its previous history: that the 'electron proper, namely the scalar distribution ρ, is the focus of the whole, the essential focal property being that the field at any instant is completely determined by the previous history of the focus and of its space relations through all previous time'.[2] Further, the influence of the electron or its field streams away from it with finite velocity. As we have already noted, the electron as propagating such a field of force bears a marked resemblance to the creative advance of nature. This is not surprising, since he believes that the electron is nothing else than the expression of certain permanent recognizable features in this creative advance.[3]

Whitehead draws our attention to the fact that the two views of causation are here both included. He points out that the complete field within any region of space depends on the past histories of all the electrons, histories extending backwards in proportion to their distances. But on the other hand, he goes on, 'viewing the cause which effects changes on the electron within that region, it is solely that field within the region, which field is coincident with that electron both in time and in space'.[4] In other words, though the focal region or electron is dependent for its character upon all the other entities, nevertheless, if we confine ourselves to the region which it occupies, it is discerned as itself determining its own transmission of character.

[1] *Logic of Modern Physics*, p. 213. [2] AE, p. 227.
[3] PNK, cf. p. 97. [4] AE, p. 227.

Though it might be thought difficult to find a passage matching this in PR, yet in Whitehead's description of what is apparently a character of direct experience we are told, 'An actual entity is at once the product of the efficient past, and is also, in Spinoza's phrase, *causa sui*'.[1] 'There is the becoming of the datum, which is to be found in the past of the world; and there is the becoming of the immediate self from the datum. This latter becoming is the immediate actual process.'[2] The former process is, as we have seen, termed by him 'efficient causation', whilst he terms the latter 'final causation'. Though it is not claimed that there is a point-to-point identity, yet his earlier position does seem to bear some resemblance to his later more metaphysical view.

To return to Whitehead's account of physical causality. If we neglect the electron's past history and restrict ourselves to the present events 'which form the continuous series of situations of the electron', we can say as Whitehead does that the electron 'is entirely self-determined'.[3] This, despite the terminology used, is really a statement of physical determinism, that the future state of the electron is determined by the present event in which it is situated. Each event in the historical route making up the electron (i.e. the successive situations) may be said in this way to transmit its character to the events which are future relatively to it.

It is interesting to see that Whitehead uses the notion of self-determination here to refer to the way each event determines its own future development (or transmits its character into the future), very much as in dynamics the future state of a system is assumed to be determined in terms of a complete description of its present condition. We may compare this with his view in PR, where, when speaking of the event's process of development, he tells us that the 'subjective aim' is this subject itself determining its own development, so that it becomes the initial phase of a new process. It will be observed that the way Whitehead uses the notion of self-causation in PR strongly resembles his above use of self-determination; the two concepts function in similar types of context.

The reader may perhaps be surprised to find that Whitehead's doctrine of self-causation, which seems on the surface to be entirely opposed to physical determinism, is after all not so very different from it. This probably arises from the belief that it is Whitehead's notion of 'efficient causation' which corresponds to physical determinism. But what needs noting is that 'efficient causation' refers

[1] PR, p. 209. [2] PR, p. 209. [3] CN, p. 159.

here only to the past history of the system, and that 'determinism' as ordinarily used in physics neglects the historic cause, and takes the present condition of the system as determinative of its dynamics. From this point of view it therefore has much more in common with Whitehead's view of final causation.

As we have already seen, in this physical concept of fields of force is to be found the germ (if not the basis) of many of the doctrines elaborated in P R. Consider, for example, the mutual immanence of events—that every event 'has a perfectly definite bond with each item in the universe'. This resembles his view in P N K and C N, that the stream of events along which a certain character is transmitted is modified, in however slight a degree, by the character of other events throughout the rest of the universe.

On closer examination then, it seems that Whitehead's notions of 'efficient' and 'final' causation, far from conflicting with modern physics, resemble at least in some respects the causal factors operative in physical systems having the character of historicity. An example of this may be observed in the electro-magnetic field, where the future is not only determined by the present process, but also by its previous history. Such systems also share this feature with biological organisms whose behaviour can only be understood satisfactorily as a function of their total history. As we have seen in Whitehead's account of the development of an event, its initial phase is dependent upon the antecedent events (i.e. efficient causation). On the other hand, its present development is directed by the vector of the immediate process, which he chooses to term 'subjective aim' or 'final causation', much as in the case of the electron he conceived the focal region as propagating an emanation from itself at each instant.

II

Whitehead discussed the notion of fields of force in P of R under the heading of the physical field of natural science which, by limiting the contingency of sense-experience (i.e. the possibilities of future sense-experience), enables us to predict the character of future events. It is assumed, he tells us, 'that the apparent adjectives of the past indicate a certain distribution of character throughout events extending from the past into the future'.[1] Putting this in perceptual terms, we may say that e.g. the table we

[1] P of R, p. 71.

have just been looking at indicates that the stream of future events which is the continued existence of the table will manifest a certain continuity of character. If, for example, we go out for a walk we expect when we come back again into the room that we will still find the table standing there. It gives us, as it were, a rule for the prediction of future events. Or consider another perceptual example: when we see a bull coming towards us, we assume that certain kinds of experience are likely to follow on this event—that if we do not move quickly out of the bull's way, he may get uncomfortably close.

Whitehead has already indicated that an adjectival particle, by which he means a particular perceptual or scientific object (or stream of events pervaded by some such sensory or physical characters), receives its enduring stability from the individuality of the historical route which it qualifies. The whole notion, he tells us, is practically the same as Faraday's conception of the tubes of force as constituting the physical particle, with the modification that in the act of streaming through space it streams through time.[1] As far back as *The Organisation of Thought*[2] (1917) Whitehead pointed out that what in experience corresponded to the thought object of perception was really a stream of sense-presentations composed of very similar characteristics whose modifications increase only gradually. It thus formed a homogenous component stream within the composite stream of sense-experience.

Whitehead made this aspect of perceptual objects more explicit when he asserted that ultimately they are pervasive adjectives which are controls of the ingression of sense-objects.[3] They are regarded as such 'controls' in the sense that they give us a rule as to which specific characters will exhibit themselves in the immediate future; for example, the grass exhibits itself as green, the bell exhibits itself as tolling and the stone as touchable. In this way they limit the kinds of experience we can expect to follow our perception of certain types of perceptual object. This enables us, for example, when we observe the bull, to extrapolate to others of its characteristics, so that we quickly get out of the way.

As Whitehead points out, 'it thus means, in its essential character, a control of the future from the basis of the present'.[4] In other words, the perceptual object exhibiting itself in the present

[1] P of R, cf. pp. 75-6. [2] Republished as A E, cf. pp. 186-201.
[3] *Proceedings of the Aristotelian Society* 1922-3, p. 17. [4] Ibid, p. 17.

situation in the form of a particular sensory character gives us such a rule (that the character of the future event will conform to that in the present) for anticipating the appearance of the specific sensory characters in the immediate future. When, for example, we observe a coloured patch which we take to be a table, we assume that unless anything unexpected happens we shall observe a similar coloured patch when we come into the room the next day.

Whitehead now explicitly identifies a perceptual object with the notion of a field of force: 'Thus in modern scientific phraseology, a perceptual object means a present focus and a field of force streaming out into the future. This field of force represents the type of control of the future exercised by the perceptual object—which is, in fact, the perceptual object in its relation to the future, while the present focus is the perceptual object in its relation to the present. . . . What we observe is the control in action during the specious present.'[1] This identification may seem strange to the reader, since in our everyday life we usually assume that an object is a permanent entity with a definite set of qualities existing here and now. However, it is not altogether misplaced, if one accepts Whitehead's assumption that perceptual objects are really routes of events pervaded by some definite sensory character. His view is then not so far removed from Faraday's—that the physical particle is really a tube of force along which there is a transmission of physical character (except, of course, that it is formulated in perceptual terms). Whitehead does, however, pass rather abruptly from his earlier description of the stream of crude experience, whose sensory texture becomes gradually modified throughout its course of development, to the more precise notion of a transmission of sensory character along a route of events. This notion, like the conception of tubes of force, does not, at least to us, seem to be descriptive of direct experience. In both cases we seem to be dealing with abstract constructions.

Whitehead then thinks of the perceptual object after the fashion of the electron, as a focus or scalar distribution in the present situation which transmits its character into the future. In this way it regulates or controls the appearance of future sensory data. When he states that what we are aware of is the control (i.e. perceptual object) in action, he refers to the manner in which the object exhibits itself in a particular stretch of the route, i.e. the bell as tolling, the stone as touchable, etc.

[1] Ibid, p. 17.

Further, Whitehead was at pains to point out in P of R[1] that among the experiences upon which the perceptual object is based are expectations of its future history. We assume that it will go on being the same kind of object in the future, and this assumption that the future will conform to the present is hence an integral part of our notion of a perceptual object, and enables us to submit our contingent perceptions to some measure of control.

In his later writings, as we have seen, Whitehead also attempts to draw an analogy between direct experience and the concept of the electro-magnetic field. 'There is thus an analogy between the transference of energy from particular occasion to particular occasion in physical nature and the transference of affective tone, with its emotional energy, from one occasion to another in any human personality. The object-to-subject structure of human experience is reproduced in physical nature by this vector relation of particular to particular.'[2] The object-to-subject structure of human experience comes to much the same thing as his notion of the conformation of character along a route of events. Hence, when he says it is similar to the vector relation of particular to particular in physics, he means that, as in the case of physical transmission, a character is transmitted from the event in our immediate past (i.e. the object) to the event we are now experiencing in the present (i.e. the subject).

But we do not think Whitehead is justified in drawing such an analogy between the transference of physical energy and the stream of direct experience. Whitehead's doctrine of the transference of subjective form does not seem to be simply a description of experience but appears to be modelled on the notion of transmission in electrical systems with radiation. As he himself states, this process of inheritance corresponds to the physical flux in nature, or, to be more precise, Poynting's Flux of Energy in electrodynamics.[3] But, even if we grant that there are points of resemblance between the perceived passage of nature and the flux of energy described by physics, we still fail to see what similarity there is between 'emotional intensity' and 'physical energy'. The former notion simply refers to the intensity of the affective tones of the specific sensory characters in our experience. It certainly seems different from the notion of energy in physics, which in its most concrete form refers to mechanical or other physical systems in which work is performed.

[1] P of R, cf. p. 73. [2] A I, p. 242. [3] A I, cf. p. 238.

Whitehead further believes that 'The notion of physical energy, which is at the base of physics, must then be conceived as an abstraction from the complex energy, emotional and purposeful, inherent in the subjective form'.[1] However, 'energy' as used in physics refers to the work done, or capable of being done, by physical systems. On the other hand, emotional intensity does not satisfy the same criterion as the physical notion of energy, having nothing to do with the performance of (or capacity for doing) work. Hence, to speak in this sense of emotional intensity as emotional energy (as that from which the physical notion of energy is derivative, or is an abstraction) is to confuse different levels of generality. Energy seems to be a general property or function of physical systems, whereas emotional intensity is something directly experienced by us. Though the energy concept is a general property of such systems, and if considered by itself is an abstraction, one thing is clear: it certainly does not seem to be an abstraction from the affective characters in our experience. He could, of course, counter this by arguing that on his account the physical events in nature are to be thought of after the fashion of throbs of emotional energy.

Whitehead draws a more radical comparison between the elements in direct experience and the notions of physics when he states 'If we substitute the term "energy" for the concept of a quantitative emotional intensity, and the term "form of energy" for the concept of "specific form of feeling", and remember that in physics "vector" means definite transmission from elsewhere, we see that this metaphysical description of the simplest elements in the constitution of actual entities agrees absolutely with the general principles according to which the notions of modern physics are framed'.[2]

This is as complete an identification as one could hope to find. Whitehead is arguing that, just as in our experience emotional intensity is, as it were, clothed by some specific sense-datum, so in physics energy manifests itself in different forms, in the shape of electrons, photons, etc. He assumes that these physical factors differ after the fashion of sensa and their emotional intensities and may in a similar way be pictured as clothing physical energy. As he puts it, 'the sensa in metaphysics are the basis of the diversity of specific forms under which energy clothes itself'.[3]

No wonder Whitehead is able to announce that the general

[1] A I, p. 239. [2] P R, p. 163. [3] P R, p. 163.

principles of physics are exactly what we should expect as a specific exemplification of the metaphysics required by the philosophy of organism. And as if to justify this identification he goes on: 'In this way, the philosophy of organism . . . appeals to the facts'.[1] An unkind critic might at this stage remark that such notions as 'scalar localization', 'vector transmission' and 'forms of energy' represent very different kinds of fact from a bang in the eye, a man running after his train and a beautiful landscape. They seem to be complex intellectual constructions, which need not have any resemblance to the facts they order, so long as they fulfil their function in the ordering of experiential data.

However, there is an additional complication. As we have seen Whitehead believes that the activities in physical nature have similar properties to the elements in direct experience. He therefore tends in his account of the physical world to employ the same epistemological language as when describing direct experience. Thus when he sometimes speaks of 'quantitative feelings' he really has in mind physical energy, which he compares to a low-grade form of emotional intensity. Similarly, when he talks of 'qualitative feelings' or 'sensa', he often refers to particular forms of energy, since he assumes they are the primitive prototypes of the sensa directly experienced by us. The difficulty facing the reader is, of course, to ascertain the precise context in which Whitehead uses these terms. In the above passage he seems to employ them in a general sense to refer to elements both in experience and physical nature.

III

Let us examine in more detail how the concept of energy is actually used in physics. As we have noted, 'energy' defined as a capacity for work or amount of work done is descriptive of the behaviour of certain physical systems. For example, a bent spring possesses energy since it performs work when it recoils, and bodies in motion possess energy, since work is performed when they overcome obstacles. When we talk of a quantity of energy, we are dealing with something very different from a degree of emotional intensity, with the way physical systems behave when certain changes are produced in them; how, for example, a charge of gunpowder explodes when a match is applied to it.

[1] P R, p. 164.

It is therefore pertinent to ask, 'how does "emotional intensity" correspond to a capacity for performing work (or work performed), and how do specific sensa, such as colours, sounds and tastes, correspond to the specific forms of energy—motion, heat, electricity, etc., which are capable of performing mechanical work?' Energy does not seem to be an observable quality belonging to an object, like its particular colour or shape, nor for that matter does it seem to be a Gestalt quality. Apart from specific forms of work performed there is nothing else observable. Consider the relationship between a specific form of energy, say heat, and the notion of energy itself. We can only measure the quantity of energy it possesses (i.e. the quantity of work it is capable of performing) when it is made to do some work such as moving a piston.

The fallacy inherent in Whitehead's view is that he assumes that energy is something like a substratum, and is clothed by the specific forms of energy in the same way as sensa are assumed to clothe emotional intensities. This is, however, to reify the energy concept, transforming what is a general property or function of a whole physical system into a particular characteristic. Further, the relation between the specific forms of energy and the general energy concept is rather like that between ϕa and $\phi \hat{x}$, i.e. between an observable specific physical system (performing or capable of performing work) and an abstract formula, which is applicable to any kind of physical system and to any type of work performed.

In its abstract form the energy concept is a purely functional relationship between mass, length and time, a rule by means of which we can calculate the capacity of physical systems for performing work. 'Energy is particularly simple, in that it is connected with measurable properties of the system by a simple formula ($\Sigma \frac{1}{2} mv^2$).'[1] Since energy is then merely a function of certain measurable properties of physical systems, Whitehead has little justification in giving it a qualitative content.

Against this background, consider Whitehead's statement that 'The science of physics conceives a natural occasion as a locus of energy. . . . The words electron, proton, photon . . . all point to the fact that physical science recognizes qualitative differences between occasions in respect to the way in which each occasion entertains its energy.'[2] But such qualitative differences on analysis resolve themselves into quantitative ones. On the purely physical plane, the qualitative experiences of, for example, heat and light

[1] Bridgman, *Logic of Modern Physics*, p. 112. [2] A I, pp. 237-8.

are excluded, and in their place we substitute thermometer and photometer readings so that we merely deal with abstract sets of quantities. These physical qualitative differences (i.e. diverse forms of energy) which he tells us 'are entirely constituted by the physical flux of energy' turn out then, as long as Whitehead talks physics, to be still quantitative in character. Hence, the analogy Whitehead draws between 'forms of energy' in physics and emotionally toned sense-qualities breaks down, and with it his thesis that there is a straightforward one-one correspondence between direct experience and the events in the physical world.

Whitehead in his account would seem really to have in mind a specific kind of energy, namely electrical energy, so that he can speak of radiation within an electrical system in terms of the transference of throbs of emotional energy. It is true that in ordinary conversation we often speak of a situation as being tense or electrically charged when something dramatic occurs. It may be that Whitehead has been misled by such metaphorical ways of speaking, but the analogy between electrical energy and emotional intensity soon loses its plausibility when one examines the energy concept and takes into account other forms of energy. There is not, for example, the slightest resemblance between 'emotional intensity' on the one hand, and a revolving wheel, a bent spring recoiling or gunpowder exploding, on the other.

It is interesting to note that Whitehead himself admitted that he arrived at his organic conception of nature not by way of psychology or physiology, but through physics. 'Mathematical physics', he tells us, 'presumes in the first place an electromagnetic field of activity pervading space and time. The laws which condition this field are nothing else than the conditions observed by the general activity of the flux of the world, as it individualizes itself in the events.'[1] After describing the transmission of an electrical charge as a flux of aspects throughout space and time, he goes on, 'We may conclude, therefore, that the organic theory represents directly what physics actually does assume respecting its ultimate entities'.[2]

However, no one has yet directly observed such a field (or physical flux of energy); all that has been done is to define it in terms of measurable charges. What the physicist actually does, as Bridgman points out, is to place an exploring charge at various points where electrical phenomena are observed, measure the force

[1] S M W, p. 190. [2] S M W, p. 193.

on it, and then proceed to extrapolate this process to every point of space. But we have little reason for supposing that the field of force thus constructed is itself really in nature, in the same way as the measurable charges. It seems merely to be a device used by us to correlate certain electrical phenomena. Only when we give the field a physical reality do we need to demonstrate that there is a structural similarity between it and direct experience.

Despite Whitehead's frequent warnings to philosophers to guard against the fallacy of misplaced concreteness, he would appear to confine his criticism to the concepts of Newtonian physics. He certainly seems to be giving the electro-magnetic field a form of reality, in the same way as the philosophers of the seventeenth and eighteenth centuries gave a physical reality to the physical concepts of that period. And, as in the case of classical physics, there is a broad similarity between the modern physical concepts referred to above and direct experience. But it is quite another matter to go on to suppose, as Whitehead does, that this correspondence is more than a very rough and ready one, and that his philosophy enables him to 'find in descriptions of human experience factors which also enter into the descriptions of less specialized natural occurrences'.[1] In this way he hopes to 'point out the identical elements connecting human experience with physical science'.[2] The fact is that, though they may possess identical elements, the similarity between them is usually of an extremely attenuated nature.

As a final piece of evidence, let us look again at what Whitehead meant by the 'physical flux of energy'. As we have seen, he uses this concept to refer to the way events inherit their energy from past events and transmit it into the future. After identifying this transmission with the Poynting Flux of Energy in electro-dynamics, Whitehead confesses that he first heard of it when as a young graduate student he listened to a lecture delivered by Sir J. J. Thomson. And he goes on to say: 'The sole conclusion with which we are concerned is that energy has recognizable paths through time and space. Energy passes from particular occasion to particular occasion. At each point there is a flux, with a quantitative flow and a definite direction.'[3] He then proceeds to draw an analogy between the physical flux and what he terms the vector character of experience.

Time seems to have taken its revenge. Both Poynting and

[1] A I, p. 237. [2] A I, p. 237. [3] A I, p. 238.

Thomson were inclined to believe in the physical reality of the lines of force in the electro-magnetic field. Thomson, for example, stressed the need for bringing home to the student of electro-dynamics the physical reality of the tubes of force; that they should not take Maxwell's theory as merely being the solution of certain differential equations, but as referring to actual physical events in nature.[1] The tendency in physics today is in the reverse direction, rather to take the electro-magnetic field as a piece of symbolic machinery or a pragmatic device. As Bridgman says, 'I believe that a critical examination will show that the ascription of physical reality to the electrical field is entirely without justification'.[2] In any case, whether we take up a realist position or not, our other criticism stands, that physical descriptions and descriptions of experience are not isomorphic.

SUMMARY

It might be argued that Whitehead's postulation of two types of causal process conflicts with the notions of physics, as for Whitehead an actual entity is at once the product of its past and is also self-determinative. This doctrine is, however, not opposed to physical determinism, which takes the present condition of the system as determinative of its dynamics; and further, in electrical systems with radiation, the past history of the system has also to be taken into account.

The analogy Whitehead draws between forms of energy in physics and emotionally toned sensory qualities fails. What he would seem to be doing is to reify the electro-magnetic field in the same way as philosophers of the seventeenth and eighteenth centuries gave a physical reality to the physical concepts of their period.

[1] *Notes on Recent Researches in Electricity and Magnetism*, 1893, p. v.

[2] *Logic of Modern Physics*, p. 57. It has also been pointed out that the attempt to ascribe reality to energy flow in the electro-magnetic field leads to absurdities and contradictions; that with a stream of water we can say a particle is now here, now there, with energy it is not so. Poynting's Flux of Energy is therefore to be taken as a mere mathematical abstraction. (Cf. Sir James Jeans, *The Mysterious Universe*, pp. 108-9.)

CHAPTER XVI

PHYSICAL THEORY IN THE *PRINCIPLE OF RELATIVITY*

AS we have seen, Whitehead's view that physical feelings are the atomic bricks out of which all our physical relationships are constituted[1] is not new to P R. It was elaborated to some extent in P of R. He there spoke of the field of an adjectival particle *m* (physical or perceptual object) in the present situation *P* as a limited atomic region stretching from *P* into *P*'s futurity, and pervaded by a character due to *m*. Whitehead conceives the object *m* as the present focus of the pervaded historical route or field stretching from the past into the present. The atomic field stretching from *P* he called *P*'s causal future.[2] As will readily be seen, he was merely putting forward here the notion of the transmission of physical character from event to event.

Though, for ordinary purposes, the object *m* may be treated as being simply located in the present event, Whitehead believes that both in modern science and in epistemology the relevant objects need to be considered as having a temporal spread requiring a period to manifest themselves. For Whitehead, therefore, they have to be thought of as historical routes of events pervaded by definite patterns. On this view, the present focus (sensory or physical pattern characterizing the present event) is, as it were, an instantaneous three-dimensional section of the field as it streams from the past into the future. The field and its focus, namely the route, and that section which we term the present object, are not to be thought of as independent objects, but as correlated in one organized unity. The whole field streaming from the past into the future is the object.

This is the forerunner of Whitehead's discussion of physical feelings in P R. For the most part these respective accounts run parallel, except that in P of R the discussion is in terms of event-particles instead of events. The present situation *P* corresponds to the 'subject' or 'present event'. The limited atomic field stretching

[1] P R, cf. p. 335. [2] P of R, cf. pp. 34-6 and pp. 71-6.

into the future from the present, and pervaded by a definite character (due to the object *m* at *P*), corresponds to the 'simple physical feeling' entertained by the subject (i.e. the transmission of character from the present into the future). The object *m* or present focus of the field corresponds to the 'objective datum'—the character transmitted to the present from the past, or the field stretching from the causal past of *P* into *P* itself.

We need to show in more detail how Whitehead's account of physical feelings is related to the particular kind of physical theory elaborated in P of R. In this context a 'simple physical feeling' resembles what he called an element of impetus, i.e. a stretch of a historical route having certain physical properties (e.g. mass and electrical charge). He tells us that in order to exhibit the character of the physical field due to a mass particle (or physical object) we have to consider it as pervading a 'kinematic element' of its route. By this he means a stretch of the route which, unlike an event-particle, retains the quality of historic passage. Each kinematic element such as XX' having X as initial starting point (or focus) will hence be pervaded by certain physical characters. Two such characters are the potential mass impetus of the particle (its mass or inertia), and its potential electro-magnetic impetus (or electric charge).[1]

In C N we are told that, by means of such elements of impetus (or stretches of historical routes pervaded by specific physical characters), he is able to express the field of activity of events (gravitational and electro-magnetic fields) in the neighbourhood of some definite event-particle (or point-instant) *E* of the four-dimensional manifold. 'The event-particle *E* is related to any neighbouring event-particle *P* by an element of impetus. The assemblage of all the elements of impetus relating *E* to the assemblage of event-particles in the neighbourhood of *E* expresses the character of the field of activity in the neighbourhood of *E*.'[2]

What Whitehead is talking about here are purely physical quantities, and hence his account ought not to be given an experiential content. He is referring to the assemblage of physical quantities defining the inertial and electro-magnetic properties of some definite route or kinematic element which constitutes the physical field in the neighbourhood of *E*. In other words, the physical field at an event-particle *E* is such an assemblage of physical quantities which can be defined mathematically.

[1] P of R, cf. pp. 78-80. [2] C N, p. 181.

Though Whitehead is dealing with event-particles, he yet stresses that each event-particle is related to other event-particles in the four-dimensional (physical) continuum by means of such elements of impetus or 'atomic physical fields'. They are the sets of physical quantities referring to the gravitational and electro-magnetic properties in the neighbourhood of E. The whole assemblage of such atomic fields makes up the gravitational and electro-magnetic fields in nature. On his view, then, material particles are pervaded tracks of event-particles in the space-time continuum. These tracks resemble Eddington's four-dimensional world tubes demarcated from the rest of space-time by more or less abrupt boundaries. As Whitehead considers each pervaded route to be really a field of force, its influence or physical field extends throughout the continuum.

Let us compare the above with his statement in PR that all our physical relationships are made up of physical feelings as their atomic bricks—that each event has a definite bond with each item in the universe. In other words the actual world is a manifold of prehensions, where a prehension is conceived as a concrete fact of relatedness[1] (or a stretch of transmission). When Whitehead then speaks of an event as prehending another by means of its simple physical feelings, he seems to say very much the same thing as when he speaks of event-particles as being related to each other by means of their 'atomic physical fields' (or elements of impetus). However, in the latter case his whole account is expressed in purely quantitative terms. It is stated in terms of a series of space-time measurements, expressing the quantitative aspects of the historical passage of the kinematic element XX', and certain physical quantities (physical measurements) expressing the inertial and electro-magnetic properties of this element.

This account forms the basis of Whitehead's own brand of relativity theory. He brings out his divergence from Einstein in this matter by saying, 'Where I differ from Einstein is that he conceives this quantity which I call the impetus as merely expressing the characters of the space and time to be adopted and thus ends by talking of the gravitational field expressing a curvature in the space-time manifold'.[2]

Roughly what Whitehead does is to accept a uniform (physical) space-time continuum (or Euclidean geometry) weighting it with the individual peculiarities of the gravitational and electro-magnetic

[1] PR, cf. p. 29. [2] CN, pp. 181-2.

fields. He expresses these fields in terms of physical characters (or rather physical quantities) which he conceives as pervading stretches of historical routes of event-particles making up the four-dimensional space-time continuum. As material objects for Whitehead are such pervaded routes or fields of activity, the physical peculiarities or modifications set up in them by the presence of other pervaded routes in the neighbourhood are therefore expressed by him in terms of physical causality. But instead of having material masses secluded within regions of space in which changes only occur as a result of external forces, we have on Whitehead's view four-dimensional routes pervaded by physical characters. These extend throughout the space-time continuum influencing and modifying each other.

Einstein's treatment is simpler, since by merely using spatio-temporal concepts he is able to express what for Whitehead are physical peculiarities (or rather physical quantities) in terms of curvatures in space-time (or a set of spatio-temporal measurements). In this way he short-circuits Whitehead's account, expressing by means of a Riemannian geometry what Whitehead needs to express by a Euclidean geometry and a physical field of force. Einstein conceives objects as curvatures (or ridges) within space-time (world lines or four-dimensional tubes which are the tracks of the particles). The gravitational effects in the neighbourhood of such tracks are accounted for by the way the connecting regions of different tracks bend towards each other. On Whitehead's view the ridges become such pervaded routes, and the curvature set up in the intervening space-time a field of activity streaming away from these routes. Hence Whitehead substitutes variations in the intensity of a physical field for what on Einstein's view are variations in the curvature of the space-time manifold.

Whitehead's early arguments against Einstein are bound up with his view that the elements of a physical theory need to correspond with the elements within our perceptual field, which is nearly or approximately uniform. Though, it is true, we do not observe the world around us as in a distorting mirror, there is no reason why Einstein's physical theory should not be accepted, if we take it merely as a methodological device for the correlation of certain observed phenomena.

Whitehead's belief in the uniformity of physical space-time seems, then, to be based on his assumption that the structure posited by a physical theory must exactly reproduce the structure of perceptual fact. In his argument Whitehead assumes that, because perceptual space-time is uniform, the space-time dealt with by the physicist must also have this property. But this necessity does not arise if we regard physical space-time as largely a device for correlating certain physical characteristics.

SUMMARY

The forerunner of Whitehead's philosophical view that physical feelings are the atomic bricks out of which our physical relationships are constituted is to be found in P of R. In this context, a simple physical feeling would seem to resemble what he called an element of impetus—a stretch of a historical route having certain physical properties, i.e. inertial and electro-magnetic properties.

On Whitehead's view material particles are pervaded tracks of event-particles in the spatio-temporal continuum. This conception of event-particles as related in the four-dimensional spatio-temporal continuum by such elements of impetus forms the basis of Whitehead's Theory of Relativity. Whitehead, as against Einstein, accepts a uniform physical space-time continuum, weighting it with the individual peculiarities of the gravitational and electro-magnetic fields. But there seems little justification for arguing, as Whitehead does, from the supposed uniformity of our perceptual field to the uniformity of physical space-time.

CHAPTER XVII

DETERMINISM AND FREE WILL

I

WHITEHEAD, after an appeal to the Romantic poets, criticizes the mechanistic theory of the eighteenth century which extruded the volitional and purposive elements from nature, and led to the doctrine of determinism—the belief that our behaviour is completely determined by physical laws. He summarized this view briefly as follows: 'Each molecule blindly runs. The human body is a collection of molecules. Therefore, the human body blindly runs, and therefore there can be no individual responsibility for the actions of the body.'[1] But, as Whitehead points out, such a view conflicts with our direct experience of free will, since we are able to effect changes in our bodily states; for example, if we wish to move an arm or turn our heads, we can do so. And if this is the case, he argues, the molecules in our body do not blindly run and mental states have some relevance when we discuss our behaviour.

What should be noted (though Whitehead is not very explicit on this point) is that he is not arguing against the classical scheme of physics, but against the philosophers who took it over, and gave it a concreteness it did not possess. If one recognizes the scheme for what it is, a method of representing certain quantitative aspects of nature, then none of the above difficulties arise. The fact that this theory did not include our moral intuitions is no argument against it, for they were explicitly excluded. Further, the concepts of classical physics have been superseded not because they did not include 'that fundamental intuition of mankind which finds its expression in poetry',[2] but because another method of representation was found which fitted the physical facts better. Even if the physicist does substitute the notion of energy for matter at an instant, he is merely substituting one way of representing certain quantitative aspects of nature for another. The ordinary man in the street has not noticed any change in his *Weltanschauung*

[1] S M W, p. 96. [2] S M W, p. 119.

since physicists have swung over from classical mechanics to field physics and quantum mechanics.

Whitehead, as we have seen, holds that nature has the full content of our perceptual experience, in which he includes our moral and aesthetic intuitions. He therefore believes that the doctrine of materialism only applies to very abstract entities, the product of our logical discernment. On his view, however, the concrete enduring entities (i.e. biological as well as physical objects) are organisms, so that the plan of the whole influences the character of the various subordinate organisms even down to the smallest electron. Hence, 'an electron within a living body is different from an electron outside it, by reason of the plan of the body . . . and this plan includes the mental state. But the principle of modification is perfectly general throughout nature, and represents no property peculiar to living bodies.'[1]

What Whitehead seems to be doing here is to substitute for materialistic mechanism the field theory of modern physics and call it 'the theory of organic mechanism'. It therefore can be said in no sense of the word to conflict with physical science. In S M W he appears to push this theory as far as he can in his descriptions of direct experience and physical nature. As he states elsewhere, 'The fundamental fact, according to the physics of the present day, is that the environment with its peculiarities seeps into the group-agitation which we term matter, and the group-agitations extend their character to the environment'.[2] On this view there are no self-contained things within limited regions of space, since their influence streams throughout the utmost recesses of space and time, i.e. they are fields of force. Of course physical theory has moved since Whitehead wrote, but it is this unrestricted application of the field theory which gives Whitehead's philosophy a Hegelian tinge. However, it is questionable whether the notion of organic mechanism is in any way more concrete than the deposed materialistic mechanism, since the fundamental character of physics is that it is abstract (or quantitative) and excludes all qualitative detail.

Whitehead puts forward the theory of organic mechanism (or rather the field theory) in the following terms: 'it remains', he tells us, 'an immediate deduction that an individual entity, whose own life-history is a part within the life-history of some larger, deeper, more complete pattern, is liable to have aspects of that

[1] S M W, p. 99. [2] N L, pp. 31-2.

larger pattern dominating its own being, and to experience modifications of that larger pattern reflected in itself as modifications of its own being. This is the theory of organic mechanism.'[1] To give a simple physical illustration, in the case of an electromagnet the molecules within the iron core will mutually attract and repel each other in accordance with the direction of the current flowing through the surrounding coil. Nevertheless, the theory remains an immediate deduction only if one accepts as the premise of the above argument the 'field' with its speculative superstructure. It might be possible to explain the change in the molecules by 'action at a distance' theories without assuming any intermediary field.

If we take the human body as such a patterned field of activities, then since the molecules within it are themselves fields of activity, they will behave very differently from the ones outside it. Hence, the influence of the larger bodily pattern in which they function cannot be ignored. In other words, the molecules within living bodies will exhibit characteristics not to be observed amid inorganic surroundings. Developing this theme, he goes on, 'The prompt self-preservative actions of living bodies, and our experience of the physical actions of our bodies following the determinations of will, suggest the modification of molecules in the body as the result of the total pattern'.[2] Or, putting it rather differently, our experience of the volitional control we have over our bodies suggests that the included molecules are modified by volitions, since the latter also form part of the total field (or experienced body) in which the molecules are sub-patterns.

Whitehead explains how the modification of the total pattern (or body) by the determinations of will could at the same time modify the character of the subordinate patterns, for example, the molecules within it: 'the modification of total pattern would transmit itself by means of a series of modifications of a descending series of parts, so that finally the modification of the cell changes its aspect in the molecule, thus effecting a corresponding alteration in the molecule—or in some subtler entity'.[3] In other words, the changes set up in the bodily state by volition would be progressively transmitted downwards to the smallest bodily parts, for example, individual cells and molecules, though, he admits, the individual effects of volition may at this stage be quite negligible.

[1] S M W, p. 134. [2] S M W, p. 186.
[3] S M W, p. 186.

By means of this theory of organic mechanism, Whitehead believes he has secured that either determinism or free will shall have some relevance, unhampered by the difficulties introduced by materialistic mechanism; 'In this theory, the molecules may blindly run in accordance with the general laws, but the molecules differ in their intrinsic characters according to the general organic plans of the situations in which they find themselves'.[1] One thing should be noted. Whitehead is not denying that molecules follow physical laws. He is only stating that they may be so modified in their character by the total bodily pattern that they will follow slightly different laws than they would in an inorganic environment, just as a piece of iron will behave differently in a magnetic field. This is possible, since Whitehead conceives physical entities not as simply located bits of material, but as fields of force mutually influencing each other. In the case of our bodies he assumes that the total bodily pattern is a complex assemblage of such fields pervaded by diverse characteristics (which make us term them molecules, cells, etc.) co-ordinated together to form this larger pattern.

The human body then, on Whitehead's view, is to be thought of as a hierarchy of patterns of successive grades of complexity, analysable into subordinate patterns until we reach the simplest patterns (components) such as electrons, etc. As he puts it, 'men are societies of cells, and of blood, and of bones, together with the dominant society of personal human experience, and cells are societies of smaller physical entities such as protons'.[2] In the case of the human body, there is a central control on the part of our volitional experience which modifies its behaviour and hence the electrons within it. Apparently, this modification takes place as a result of the transmission of control through the nerves, cells, etc., down to the individual electrons. Any change thus set up in the body as the result of volition will be propagated downwards to its smallest component elements.

Whitehead's discussion of his theory of organic mechanism is open to serious objections. Despite his assertion that the metaphysical status of molecules is irrelevant to his theory, he overlooks that he has admitted (at least in his earlier work) that molecules and electrons are constructions and hence on a different plane from our direct observations. One thing is clear; whatever their ontological status, they are on a much more general level than the

[1] S M W, p. 99. [2] A I, p. 264.

elements given in sense-perception; for example, formulae applicable to one electron will apply equally well to any other electron, whereas our observations do not seem to have this universal character.

Hence, when Whitehead supposes that our psychological experiences, for example volitions, and the molecules of the physicist are combined in one pattern to modify each other, he would seem to commit the fallacy he has crusaded against in his earlier work, that of misplaced concreteness. We may well ask how volitions can modify such entities which have an entirely different sort of relationship to the somatically experienced body from our observations of, say, the fingers with which we write. On the other hand, how do physical entities modify such perceptual data? To suppose as Whitehead does that our volitions modify the bodily molecules is equivalent to supposing, admittedly on a more phenomenalist interpretation, that theoretical constructions can causally interact with our direct experience. But even on a realist view there would be difficulty in showing how our volitions could enter into relationship with physical entities. Whitehead's solution, as we have seen, is to assert that such entities are basically identical with experiential elements.

II

The reason for Whitehead's difficulties is that, in the above account, he confuses different kinds of description of the human organism; (i) the body directly given in sense-experience—having a somatic feeling core—and bounded by coloured external parts, our hands and feet, etc., over which we feel we have some volitional control; (ii) what for convenience we may term the physiological aspect, i.e. the system dealt with by the physiologist, who attempts to describe the scientific events which supposedly occur when we have such volitional experiences: changes in blood pressure, nerve impulses, ductless gland secretions, etc.—what a hypothetical observer would presumably observe if he looked into us. This account would not include our personal experiences; the body thus described would have the same impersonal character as, for example, the table and chairs directly in front of us. However, what the physiologist would really be considering would be the standardized (or generalized) type of organism discussed and set out in text books of physiology. This resolves itself into a representation of certain general physiological facts, gathered from the observations of a large number of investigators and put into the form of an interconnected system of general propositions. We are thus given a kind of map which brings out in relief or represents the common features of a large number of observed physiological facts. But there is this important difference. Unlike a map, the system does not directly correspond to any one set of particular bodily experiences; it merely deals with some of its general features. Hence statements about the physiological system are expressions of a higher grade of generality than those about our perceptual observations.

(iii) The physical aspect. We only concern ourselves here with certain physical aspects which the human organism has in common with other (inorganic) elements in the world, stones, rocks, houses, etc. These aspects may be regarded as general characters pertaining to all varieties of objects and are expressed in terms of molecules and electrons, or whatever the current physical conception be, functioning according to physical laws. If you prefer the terminology used by Whitehead in his earlier writings, they are ideal limits to processes started in perception (constructs based on conceptual limits) and as such have no precise perceptual equivalent. Hence, in this account our organic and sensory perceptions would be replaced by an assemblage of conceptual entities. Or if one prefers the more modern view, it becomes a scheme of symbols connected by mathematical equations. In any case, one thing is clear: the body dealt with here is so abstract that it is unlike anything we ever meet with in our normal day-to-day lives.

We need to emphasize again that in both (ii) and (iii) we deal with general systems and we cannot therefore say the same thing about them as we do about our particular perceptions. Hence Whitehead's demand for science to become more concrete overlooks its essential character, namely, that it deals, to use philosophical jargon, with the universal. Its job is not to be particular and concrete, but general and abstract. As physical science, for example, progresses, it seems to move more and more away from the world in which the poet finds so much delight.

This does not mean that important sets of correspondences cannot be set up between perceptual data and the elements in the physiological and physical systems. We know as a matter of fact that it is possible to do this. When we have a toothache, the dentist finds a decaying tooth; when we feel giddy and see spots before our eyes, the doctor may find a rise in our blood pressure, or the bio-chemist changes in the chemical composition of our blood.

No doubt the success we have had in making a sufficiently good analysis of our observations into such elements (i.e. cells or molecules) may easily have led to the belief that a complete physiological or physical account is possible, and to the view that statements about perceptual facts (or sense-data) are in principle deducible from statements about physiological and physical data. However, such a programme, when critically examined, turns out to be an attempt to deduce the more concrete and particular from the more abstract and general—a task clearly beyond us.

It may be that the molecules within the physical scheme (expressing certain physical aspects of the human organism) will follow slightly different laws from a molecule in some other system representing inorganic data. But this does not mean that volitions enter into this scheme and modify it. We are not on this plane concerned with volitions or any other psychological characteristics. There may be a certain correspondence between peculiarities in the behaviour of the molecules in the physical system describing our bodily functioning, and our perceived bodily behaviour and psychological experiences as modified by volition. But have we any right to transform such a correspondence into a causal relationship?

When Whitehead tells us that our psychological experiences enter into the plan of the total organism, he is really referring to the organism as an experienced unit in our perceptual field, though it is not denied that we are aware that we have some measure of volitional control. Whitehead thus supposes that these perceptual and organic experiences in some way causally modify the elements in the physical system. However, whatever relationship exists between them it is much less straightforward than the causal account assumes.

As a result of failing to distinguish the different ways in which the term 'body' functions in different contexts, Whitehead assumes that there is a continuity between psychological data and electrons and molecules, so that the former can set up changes (by means of transmission) in the latter. And by doing this he overlooks the fundamental difference between them, that, no matter whether you take molecules as bits of matter or 'organisms', they still possess a high degree of generality. There is a radical difference between perceptual data and scientific events, which is not merely one of degree but of kind.

It will be seen that the disjunction between free will and determinism is brought about by confusing the field of direct experience in which we are aware of ourselves as having some control over our experienced bodies, and that of science which tries to deal with behaviour in purely general terms. Only in the latter sphere does the doctrine of determinism apply, as it is a property of physical systems which accept the principle of continuity in nature. Even if it could be shown that one could deduce the future behaviour of the human body from very general laws expressed mathematically, one would still be only referring to a system of equations dealt with by the mathematical physicist. All that it would mean would be that we had discovered that certain correspondences could be set up between this system and perceptual data. By further assuming that extrapolations from this scheme will correspond to future observations, we are enabled to use it as a technique for forecasting our future behaviour. Only when we accept the physical model theory—that there must be a one-one correspondence between the elements in the mathematical system and the elements in our field of experience—are we committed to the assumption of determinism in the sphere of human action.

When Whitehead speaks about the plan of the total organism as modifying the subordinate organisms, he gets himself into difficulties. He seems to imagine that we are dealing, as it were, with a set of Chinese boxes, each being a pattern or field superimposed on the other, so that a change set up in the outside field will be transmitted to every included field. Even if we accept the field theory, he would, however, at the most only be referring to overlapping physical fields. Whitehead has no justification for starting from an account formulated in psychological language and then switching over to the physiological and physical schemes at his pleasure, supposing that it really does not matter since they are all causally connected together in the bodily organism.

It is against such a speculative background that Whitehead makes his statement that 'The prompt self-preservative actions of living bodies, and our experience of the physical actions of our bodies following the determinations of will, suggest the modification of molecules in the body as the result of the total pattern'.[1]

[1] S M W, p. 186.

We question it. It merely suggests such a modification if we assume that our psychological experiences act in the same way upon the molecule as a magnetic field does upon a piece of iron, that the molecule is in some way embedded in our experience in the same way as the iron is in the magnetic field—an analogy which is certainly misleading.

The relationship is entirely different. In the latter case, we are dealing with an observable object which enters into direct relationship with a set of measurable forces in terms of which the field is defined (i.e. we can observe the deflection of the iron). In the former case, though we can directly experience our arms, legs, etc., the molecule is not at all experienced in this sort of way, since owing to its generality that conception refers not to one particular phenomenon, but to a whole range of phenomena. By the use of such an analogy, Whitehead is led to the extreme position of having to assume that because we are aware that we can modify our actions at will, that we can move our arms and legs etc., then such scientific objects can also be modified in this way.

When Whitehead supposes that man is a complex society of cells, blood, bones, electrons, etc., a structured society in which these diverse entities are co-ordinated to give rise to man with his stream of personal experiences which dominates and controls the whole structured pattern, he falls a victim of his own fallacy of misplaced concreteness. He thereby assumes that our actual psychological experiences can causally modify such inorganic and physiological entities—though, it is true, he is ascribing concreteness not to the scientific entities of the eighteenth century but to those of the late nineteenth century.

We find him, for example, saying 'The body is that portion of nature with which each moment of human experience intimately co-operates. There is an inflow and outflow of factors between the bodily actuality and the human experience, so that each shares in the existence of the other.'[1] To prevent us from misunderstanding him, and interpreting him on purely phenomenalistic lines, he goes on to amplify this: 'Ordinary language, and the sciences of physiology and psychology, supply the evidence . . . the body supplies the basis of emotional and sensory activities, and the agitations of human experience pass into subsequent bodily functionings'.[2]

[1] M T, p. 157. [2] M T, p. 157.

In other words, the physiological functionings of the body are so co-ordinated with the functionings of human experience that there is a transfer of types of agitation. Certain stimuli from the physical world are transmitted through our bodies to give rise to specific kinds of sense-experience. Conversely, the volitions affect our body, and after a somewhat lengthy journey modify the molecules. But we may ask, how can such an interchange take place? How can volitions flow into the physiological body, and the impulses described by the physiologist into sense-experience? It has not been for want of trying that neither physiology nor psychology has as yet succeeded in showing how this supposed interchange of agitations takes place.

But, to do Whitehead justice, we must point out that the view that there can be a causal connection between direct experience and scientific events is not unreasonable, if one accepts the premises from which he starts. Whitehead thinks such a causal connection is possible, because scientific objects have been transformed into fields of force; and he tries to show that the fundamental notions of modern physics, for example, energy fields, approach close to the basic elements in human experience conceived as structures of activity. Hence, the disjunction between matter and mind which found such an important place in the older philosophies does not arise in Whitehead's philosophy. In its place he substitutes types of patterned activities, or events of higher and lower grades. There is thus for him a radical similarity between the character of direct experience and the activities investigated by physics.

The difficulties in the way of explaining how our sensations arise from physical stimuli and physiological excitations are instantly removed. Somewhere in the body a process of transmutation takes place, which transmutes the physiological patterns by stepping them up into a higher grade. (Indeed he compares the body to a complex amplifier.) In a similar fashion other physiological states give rise to our personal experiences, which apparently are able by some creative feat to originate fresh experiences, for example volitions, which flow back again through the physiological body down to its inorganic parts. This is Whitehead's more concrete rendering of the facts of nature, which would solve a great many problems if it were only true.

Our contention is that, though Whitehead has substituted the notion of the field (or pattern of activity) for matter at an instant, he is still dealing with theoretical constructions, which in many ways are far more removed from direct perceptual experience than the notion of matter which kept up some pretence of being perceived. Whitehead confuses these two types of activity, that immediately experienced by us, for example our kinaesthetic sensations which arise when we push against objects, and the physical structures of activity (or vectors) which are really algebraic patterns. As we have seen, in mathematical physics the notion of energy or activity merely appears as a certain mathematical formula ($\Sigma \frac{1}{2} mv^2$) in the system of equations referring to certain measurable quantities. Hence, any attempt to show that it is similar to our kinaesthetic sensations, or even the direct perceptual experience of activity, is doomed to failure.

Take his doctrine of organism. It is questionable whether this notion of unities functioning and having a spatio-temporal extension is, in the field of physics, as concrete as he makes it out to be. When we get down to the roots of this conception, it would seem to enter into the equations in the form of a time component, an algebraic quantity which is as far removed from our direct experience of passage as is the notion of matter at an instant. In the case of direct experience, no one usually perceives such unities with spatio-temporal extension, since our sense-data tend to blur into each other, so that it is difficult to point out just where one event begins and the other ends. A certain amount of arbitrariness thus enters into our descriptions of sense-data when we apply the term 'unities' or 'organisms' to them.

III

Whitehead's attempt to show that either free will or determinism shall have some relevance in his philosophy is then based on a mistake. Roughly, Whitehead's present view is the bifurcation theory in reverse. You suppose that electrons are in perceived nature, in the same way as the red glow of the sunset, or are in our experienced body in the same way as our volitions. This position is not new: he told us, for example, 'the molecule is really in the event in the same sense as the grin is really on the cat's face'.[1] However, the molecule is not related to the perceived event in just this simple way, as it has tied up in it the observations and thoughts of numerous investigators.

[1] CN, p. 171.

Even if the molecule is not a theoretical construction but a real particle, there would still be an enormous gap between it and our ordinary range of experiences, which would require the introduction of intervening variables. It might be contended that Whitehead's doctrine of 'causal efficacy' comes under this heading, despite his claim that it gives us direct awareness of the physical world.

Consider his assertion: 'If you once accept that the molecule is definitely determined to be what it is, independently of any determination by reason of the total organism of the body, and if you further admit that the blind run is settled by the general mechanical laws, there can be no escape from this conclusion',[1] namely, determinism. But we do not see how one can escape from this conclusion in the sphere of physics. Even if we admit that the molecule is determined by the whole physical field or scheme, then it is determined according to the laws peculiar to the field (by the laws of electro-dynamics rather than mechanics). For example, we have to take into account the past history of the system and its distribution throughout space and time, instead of merely its present condition in some definite region. But this would not mean that we had done away with determinism. On the contrary, the future state of the molecule could be calculated from a knowledge of its past and present history within that system. The field theory as used in physics offers no escape from determinism, not even if we call it organic mechanism. The reader may perhaps think that Whitehead is putting forward an alternative to determinism. This obtains its plausibility because he does not clearly bring out the difference between the determinism involved in materialistic mechanism (the giving of physical reality to the classical scheme of physics) and determinism in the sense of the deductive machinery involved in a physical theory. In this sense of the term, Whitehead's theory of organism is certainly deterministic.

As we have noted, Whitehead does not deny that the molecules within the human body run according to general laws, but holds that the laws they follow would be somewhat different from those they would follow in inorganic surroundings: 'It seems possible that there may be physical laws expressing the modification of the ultimate basic organisms when they form part of higher organisms with adequate compactness of pattern.'[2]

[1] S M W, p. 96. [2] S M W, p. 186.

Now Whitehead cannot cut himself adrift from the consequences of such a view, since, if the molecules are combined together with the volitions in one pattern, and if he admits there is a causal connection between them, they should both influence each other. The notion of transmission is a two-edged weapon. If the volitions transmit their modifications to the molecules, we may also assume that the molecules will reciprocate and modify the volitions. Thus we find ourselves in a paradox.

The molecules blindly run in the human body according to certain general laws. They transmit their character and hence determination to the rest of the body, and consequently to the included volitions (which are parts of the total pattern). It follows therefore that the volitions should be determined by these general laws, so that the argument for free will breaks down. How, for example, can we say that we freely move our arm when not only this movement, but also our psychological experiences, volitions, etc., are in the final analysis determined by the general laws followed by the molecules? But on the other hand, since our psychological experiences are part of the total bodily pattern, they should also modify the behaviour of molecules in this same pattern. The molecules cannot therefore be said to be determined by the general physical laws, since they are modified by our volitions which enable us to exercise individual control over parts of our body. Hence free will should have some relevance, and the molecules should not blindly run.

Whitehead's attempt to bridge the gap between matter and mind seems to have failed. If one set of patterns in the system follows physical laws and the other set is purely self-determinative, how can there be any action or reaction between them without each obtaining some of the properties of the other? This they cannot do without losing their essential characteristics. Imagine what confusion would result were a little freedom injected into the general laws of physics, so that molecules from a liver cell could wander off at their own free will into some brain cell. As against this, it might be argued that physics nowadays employs statistical rather than causal laws, and hence is fundamentally indeterministic. However, it is questionable whether statistical indeterminacy has any relevance to free will.

SUMMARY

Whitehead criticizes the mechanistic theory of the eighteenth century which extruded the volitional and purposive elements from nature and led to the doctrine of determinism. On his view, which seems modelled on the field theory of modern physics, the concrete enduring entities (biological as well as physical objects) are organisms, so that the plan of the whole influences the character of its subordinate parts. For example, the human body is to be thought of as a hierarchy or pattern of successive grades of complexity. Thus any change set up in the body as the result of volition will be propagated downwards to its smallest component elements.

Scientific objects have, however, a more general character than our sense-perceptions and intuitions. It is therefore difficult to see how volitions as directly experienced data could modify the bodily molecules. Whitehead in his account seems to confuse descriptions having different degrees of generality with each other, and this gives rise to paradoxes.

CONCLUSION

WHAT of Whitehead's philosophy and future philosophical thinking? On the surface Whitehead's work would seem to have little interest for the present generation of philosophers, who are rather concerned with unravelling the nature of ordinary discourse and clarifying the meaning of words, and for whom philosophical systems are anathema. However, there are a number of developments in psychology and the philosophy of science which follow what we may call Whiteheadian lines—Whitehead's emphasis on the organic character of natural systems, and his attempt to bring physics and biology closer together, have a good deal in common with the Gestalt approach to mind and nature.

From the point of view of methodology, Whitehead's suggestion that we should use axiom systems (or mathematical models) to help us in our elucidations of natural phenomena has found echoes in the new discipline of cybernetics. It is also interesting to note that the attempt to endow nature with subjective characteristics, or at least to describe it in mentalistic terms, presents features which both the Whiteheadian and cybernetic approaches seem to have in common. There is, however, this radical difference. In spite of their mentalistic terminology, cyberneticians tend to downgrade human experience to the level of inanimate mechanisms, whereas Whitehead rather tends to upgrade inanimate material to the level of rudimentary experience.

Whitehead's championing of the use of mathematical models in philosophy and science places him as a precursor of an approach which is today obtaining increasing favour in the biological and psychological sciences. Further, his conception of objects as societies of events, and his view that the order underlying the universe has a statistical character, also fit in with cybernetic doctrines. And for that matter so does his emphasis on the goal-directed character of events, and his account of natural systems as functional wholes. Further, the notion of isomorphism which runs through a good deal of Whitehead's work has played an important part in Gestalt as well as cybernetic studies.

As far as Whitehead's early works are concerned, they are still worthy of study. Their relative neglect today probably stems from the current view that philosophy is not about things, but only about language, as if language were not a natural phenomenon too. Some study of Whitehead's Method of Extensive Abstraction, in which he attempts to show how geometrical notions may be related to our inaccurate, ill-adjusted field of experience, should be of value to the philosophy student interested in the application of logico-mathematical methods to philosophical problems, even if it does, as we have seen, give rise to substantial difficulties. Further, Whitehead's criticism of the Humean approach to perception in terms of static sense-data, and his emphasis on the dynamic character of mind and nature, may yet have something to be said for them when some readjustment of philosophical views is made.

APPENDIX A

PROBABILITY

I

WHITEHEAD'S only discussion of probability seems to be that which appears in P R.[1] Nevertheless, he appears to have held definite views on this topic. Whitehead refers us to Keynes for an account of the probability theory he held at one time. According to Keynes,[2] Whitehead put forward a generalized frequency theory (or truth-frequency theory) which had the following characteristics.

(1) On this theory probability is not taken as identical with statistical frequency, although it holds that all probabilities must be based on statements of frequency, and can be defined in terms of them.

(2) The subject matter of probability is propositions rather than events. It thus includes all induction and all other cases in which we believe there are rational grounds for preferring one alternative out of a set of alternatives, none of which are certain.

(3) A proposition can be a member of many distinct classes of proposition. Each class is constituted by the particular resemblance existing between its members. To give a meaning to a statement that a proposition has such and such a probability, we must therefore specify the class to which the proposition is referred, and this is determined by the evidence at our disposal. We may know precisely or within defined limits what proportion of this class of propositions is true—which is called the truth-frequency of the class. The measure of the probability of a proposition then

[1] P R, cf. pp. 282-294.

[2] Keynes, *A Treatise on Probability*, pp. 101-2. This seems to have been communicated to Keynes by Whitehead (presumably by letter). Keynes adds, however, 'it is not to be supposed that the exposition which follows represents his own opinion'. Whitehead, however, does think that Keynes has correctly expressed the views he held at that time. P R: cf. note p. 292.

depends upon its being referred to some class whose truth frequency is known within wide or narrow limits. The relative vagueness of the probabilities employed in ordinary reasoning does not then belong to the probabilities themselves, but rather to our knowledge of them.

II

Though Whitehead does not now openly champion this theory, his account of the foundations of statistics in P R still seems to have overtones of it. He there asks how statistical theories can apply to all cases to which the notion of more or less probability is habitually applied. For Whitehead, our intuition of probability originates from two distinct elements. One of them, statistical probability (the numerical ratio of favourable to unfavourable cases), is always relevant to the evidence. But since alternative grounds of evidence exist, we need, he argues, an ultimate ground which must be irrelevant to the favourableness or unfavourableness of its members. When, for example, we consider the probability of some scientific conjecture as to the internal constitution of the stars, or the future of human society after an overwhelming catastrophe, we seem to be influenced by some sort of analogy. Every forecast, according to Whitehead, gains its validity by reason of a suppressed premise in which we assume that the particular future about which our judgment is concerned will include events having a close analogy to the contemporary events. For example, we believe in the 'permanences' of nature, that this particular table in front of us will continue to exist in the future in the same way as it has in the past.

Another requirement for a ground of induction is that the number of instances which it includes be finite, since a finite ground is essential for statistical probability. If we assume that the facts with which our argument is concerned belong to a finite system, then probable knowledge can be obtained by means of an inductive argument. Random sampling, according to Whitehead, cannot help us when we try to establish statistical probability, since the whole theory of the ratios of cardinal numbers on which it depends breaks down when we deal with an infinite number of cases. He makes the interesting point that no sample is random—it has only been selected according to some special method. A finite number of samples, each selected according to a different method, will give a statistical result dependent upon the kind of method employed.

On Whitehead's view, we are enabled to evade the breakdown in reasoning when (*a*) the number of cases is infinite, and (*b*) when we are concerned with future events. He believes that every event (1) inherits characteristics from all events in its environment, and (2) occupies a finite region. Thus, though each event has a relative causal freedom from the rest of the events in nature, it nevertheless presupposes a certain relatedness to them. He thus assumes there is a natural method of sampling events, so that we obtain, as it were, a finite sample of the universe in each experience. And from this, he claims, we derive a vague intuition as to the statistical basis of the presupposed environment, which enables us to make probability judgments.

III

Whitehead[1] tells us that Keynes' account of the nature of material laws resembles his own view on the social origin of the laws of nature. Scientists, according to Keynes, do not simply assume the bare principle of the Uniformity of Nature, but something more like the superposition of small effects, which he terms the atomic character of natural law. By this he means that the material universe consists of bodies (or legal atoms) such that each of them exercises its own separate effect. A change of the universe's total state is compounded of a number of separate changes, each of which is solely due to a separate portion of the preceding state. If we start from these legally atomic units and the laws connecting them, it is possible to deduce their effects without an exhaustive knowledge of all the co-existing circumstances.[2]

However, there is a radical difference between Whitehead's position and this atomic view. Whitehead's position in some ways resembles the organic view of nature described by Keynes, in which each unit enters into different combinations and in each of them is regulated by different laws. As Whitehead put it, the molecules within a biological organism behave differently from those outside it. The organic view is regarded by Keynes as antithetical to the atomic view, since, if different wholes are subject to different laws, knowledge of part of a whole would not lead to knowledge as to its association with other parts. He thinks that this

[1] P R, cf. p. 292, note. [2] Keynes, ibid, pp. 249-50.

might make prediction impossible and induction useless. Whitehead's position seems to be an intermediate one. He could argue that we would only require some knowledge of the co-existing circumstances, namely those in the immediate neighbouring region, and that our inference would be to some extent guaranteed by the uniformity involved in space-time.

APPENDIX B

THE MULTIPLE LOCATION (OR INHERENCE) THEORY OF PERCEPTION

THE Multiple Location or Inherence Theory first made its public appearance in Broad's *Mind and its Place in Nature*, where he dealt with attempts to patch up the common-sense view of material objects, rendered questionable by such things as illusions, changes in perspective etc., by a view which holds that 'the objective constituents of some visual situations are literally spatio-temporal parts of a certain physical object, which we are said to be "seeing" '.[1] One such attempt is the 'Multiple Inherence Theory' which Broad attributes to Whitehead.[2]

Before proceeding to Broad's account of the theory a preliminary observation should be made. For Whitehead sense-objects (or sensa) are only spatio-temporal parts of physical objects by courtesy. In the first place, sense-objects are for him simple, as all other characteristics but their redness, hardness, etc., have been shorn away. By definition they therefore cannot have spatio-temporal parts. Secondly, such physical objects are constructs out of sense-objects and volumes of space at instants of time. Whitehead contends that Space and Time, i.e. definite regions of timeless space and instants of time, are not immediately given as such in the perceived field of events, but are abstractions from it. What we are aware of is a passage of events 'extending over each other'. In such a continuous stream of happenings there is nothing corresponding to the persistent objects in regions of space which common-sense supposes we directly perceive. We only come by such notions as a result of inferences drawn from a large number of experiences.

This will give some indication as to what Whitehead thought of common-sense objects, and serve as a background against which the Multiple Inherence Theory may be seen in its correct per-

[1] Broad, *Mind and its Place in Nature*, p. 160.

[2] Professor Broad calls it the Multiple Inherence Theory, Professor Price refers to it as the Multiple Location Theory.

spective. Let us now therefore turn to Broad's account: 'It has been suggested', he tells us, 'that the objective constituent of a visual situation can be regarded as a certain region of physical space which is pervaded by a certain determinate shade of colour at a certain time, *provided that* we recognize that the relation of "pervasion" is of a peculiar kind. It must not be a two-term relation, involving only the pervading colour and the pervaded region, as we commonly suppose. It must be at least a three-term relation, involving the pervading colour, the pervaded region, and another region which we might call the "region of projection" '[1] (i.e. the region usually occupied by the observer.)

We may first note that Broad's use of the term 'pervasion' is rather an unfortunate one, as it is used by Whitehead in the definite technical sense of a sense-object pervading or being located in an abstractive element such as a volume of space, already an ideal limit to a series of events (cf. Part IV, P N K). On the other hand, the relation of situation discussed by Whitehead in Part II, which no doubt Broad has in mind, emphasizes that the sense-object as given in perception is not simply located in a three-dimensional entity, but is situated in an event, 'in something going on now', and is related to other happenings now present in nature which include the bodily event (or complex of bodily experiences of the observer). Hence by the phrase 'relation of situation' Whitehead wishes to draw our attention to the fact that we are immediately aware of nature as a whole, formed of related differentiated parts. When we observe, for example, the sense-object red in some specific occurrence, it is always given together with other happenings in one complex experience. We do not usually see red by itself but as related to a variety of other characteristics in the world around us.

The reason for Whitehead's assertion that we must adopt a wider relativity than a two-termed one is clear, as what we are aware of in perception is not a multiplicity of substances characterized by qualities but a passage of related events. We need also to note that there is a great deal of difference between an event which has the characteristic of 'passage' and a region at a moment of time, which has lost it. One is a direct fact of perception, the other is its conceptual derivative. This may seem a very minor point but upon it rests Whitehead's whole theory of material objects, which for him are constructions out of such abstractive

[1] Broad, ibid, pp. 161-2.

elements and sense-objects. There is no need for Whitehead to postulate a three-termed relation of 'pervasion'. Material objects as defined by him in P N K are by their very definition two-termed: they are constructions from experience which as given is multi-termed.

Though Broad's three-termed relation might be extended to take into account the bodily event of the observer (which could be said to fill up the gap in Broad's region of projection) it is still inadequate, as it deletes the rest of the events of nature which enter into the 'relation of situation'. Broad might on the strength of his account of the theory assert that he only said it could not be less than a triadic relation, but nothing about its not being more. Nevertheless, he tells us on another page that it is a 'fundamental and indefinable relation; and it is irreducibly triadic'.[1]

But, as we have already noted, for Whitehead the relation of a sense-object to nature is a polyadic one. Every sense-object is related to every other through the events which are the terms in this relationship, so that 'the present duration is primarily marked out by . . . an interconnected display of sensa'.[2] To do Broad justice, Whitehead does however say,[3] when proceeding to give a causal account of this factual relation, that the events which enter into it can be classified into three sets not mutually exclusive. They are (i) percipient events (the bodily awareness of the observer), (ii) events which are the situations of the sense-objects, and (iii) conditioning events, the rest of the events of nature, some of which are causally more relevant than others. But this should not mislead us, as in the next paragraph he speaks of the percipient events as being included in the polyadic relation of a sense-object to nature.[4] When one then remembers that the percipient event usually occupies Broad's 'region of projection', and the event in which the sense-object is situated his 'region of pervasion', there seems little ground for Broad's assertion that Whitehead is postulating a peculiar triadic relation.

We do not think we are being unfair to Broad. Take for example the way in which he illustrates the Multiple Inherence Theory. He tells us: 'Red, on the present view, is a characteristic of such a kind that it cannot inhere in a place simply; it can only "inhere-in-a-place-from-a-place", and this relation, which needs such a

[1] Broad, ibid, p. 163.
[2] P N K, 2nd ed. Note III.
[3] Ibid., p. 84, Sect. 23.3.
[4] Ibid., p. 84, Sect. 23.4.

complex phrase to express it, is simple and unanalysable'.[1] In other words, we substitute $\phi(ax)$ for (a) where ϕ is the sense-object red, and a, x, are regions of space. But the correct formulation of Whitehead's relation of situation is, $R(\phi, \psi, \ldots X)$; where R is the inter-related structure of events, ϕ, red, and $\psi, \ldots X$, are the other objects related in this field. It will be further noted that in Broad's description there is a gap between the two places. The relation between them is purely external as they are independent regions typifying what Whitehead has called simple location. And even if worded in terms of events instead of regions, it would still only be a partial statement of the 'relation of situation', as these events (the bodily event of the observer, and the event in which red is situated) are related to the rest of the events of nature. Thus the 'place from a place' notion loses itself within a wider relatedness.

As we have already seen, in PNK Whitehead considers material objects to be constructions out of sense-objects, volumes and instants of time. To suppose, then, that he believes there are such things as material objects in the way in which common-sense or even the more sophisticated Naïve Realist does, is to overlook the definitions given in Part IV, PNK, and also his discussion in *The Organisation of Thought*. Further, if Whitehead's account was merely an attempt to modify 'Naïve Realism', what point is there in his whole criticism of material objects, of substance and quality, of Locke's primary qualities—which are just such spatio-temporal entities? When Whitehead talks about simple location, he is not criticizing the view that qualities mark out particular happenings, as they certainly do in our perceptual field, but the notion that what we have immediately given in perception are timeless regions in which visual and tactual characteristics inhere.

The belief that the notion of simple location has reference to the way in which sense-objects are located in regions, and not to the fallacy of taking these regions as concrete facts has led to the introduction of the 'Theory of Multiple Location' with its assertion that a colour has as it were a foot in two camps, i.e. it 'is there from here'. The consequences which follow from the acceptance of this theory have been elaborated by Price. 'Thus even the humblest and simplest material object, such as a table or a penny, is really a sort of infinitely various porcupine, which is not merely here in this room (as we commonly take it to be) but

[1] Broad, ibid., p. 162.

sticks out as it were in all sorts of directions, and to all sorts of distances, "from" all of which it has its being and is qualified in various ways, whether present to anyone's senses or not.'[1] For this reason it has been called the Multiple Location or Inherence Theory, as a material object is said to have its colour etc., only from an infinite number of other places or points of view.

The above picture certainly bears some resemblance to Whitehead's doctrine of perspectives. However, it wrongly assumes that by a sense-object's being multiply related Whitehead meant that this multiple relation was a feature of 'the common-sense material world' rather than of nature conceived as an inter-related structure of events. The material object with its qualities has therefore been taken, multiplied as it were in perspective, and, with all its possible perspectives added on to it, assumed to be a permanent piece of furniture of the external world.

The whole confusion then has its origin in the assumption that by 'nature' Whitehead meant the 'common nature' of the Naïve Realist. What he was really doing in his account was pointing out the happening 'there' in which the sense-object is situated, and the happening 'here' or the bodily event of the observer, two events which are pre-eminent in our awareness. Whitehead abandons the classificatory Aristotelian two-termed approach to perception with its assumption of self-subsistent entities characterized by qualities, adopting instead a functional approach, describing a sense-object as related to the whole of the happenings of nature.

[1] *Perception*, p. 56.

INDEX

For Product Safety Concerns and Information please contact our EU representative GPSR@taylorandfrancis.com
Taylor & Francis Verlag GmbH, Kaufingerstraße 24, 80331 München, Germany

www.ingramcontent.com/pod-product-compliance
Lightning Source LLC
LaVergne TN
LVHW011123300926
849160LV00008B/485
9781138871069